CAREERS IN
Renewable Energy

CAREERS IN
Renewable Energy

GET A GREEN ENERGY JOB

Gregory McNamee

PIXYJACK PRESS LLC

Careers in Renewable Energy: Get a Green Energy Job

Copyright © 2008 by Gregory McNamee

Published by PixyJack Press, LLC   PO Box 149, Masonville, CO 80541 USA

First Edition 2008

9 8 7 6 5 4 3 2 1

ISBN 978-0-9773724-3-0

Library of Congress Cataloging-in-Publication Data
McNamee, Gregory.
  Careers in renewable energy : get a green energy job / Gregory McNamee.-- 1st ed.
     p. cm.
  Includes index.
  ISBN-13: 978-0-9773724-3-0
  1. Clean energy industries--Vocational guidance. 2. Renewable energy sources. 3. Energy industries--Vocational guidance. 4. Environmental sciences--Vocational guidance. 1. Title.
  HD9502.5.C542M35 2008
  333.79'4023--dc22
                            2007050090

Printed in Canada on chlorine-free,100% postconsumer recycled paper.

ENVIRONMENTAL BENEFITS: PixyJack Press saved the following resources by printing the pages of this book on chlorine-free paper made with 100% post-consumer waste: 63 trees, 44 million BTUs of energy, 5,530 pounds of greenhouse gases, 22,952 gallons of water, and 2,947 pounds of solid waste.

Cover photos courtesy of Global Resource Options (solar array), Bonus Energy (wind turbines), and Hydrogenics (fuel cell bus).

Book design by LaVonne Ewing.

"Whatever you can do, or dream you can, begin it.
Boldness has genius, power and magic in it."

— GOETHE

Contents

continued

Introduction

Introduction

In 1987, running a side-by-side refrigerator used about 950 kilowatt hours of electricity and cost about $150 a year to run. Twenty years later, a comparable refrigerator uses half the electricity, to the benefit of both environment and wallet.

In 1975, there were 3,775,427 automobiles on the streets of Los Angeles. Today there are more than 5,200,000, but air-pollution levels have fallen in half, and an increasing number of those automobiles are hybrids or rely on biodiesel or other renewable fuels.

As long ago as the 1890s, a third of the homes in Southern California were equipped with solar water heaters. By the 1920s, the number had fallen to almost none. The number began to rise again in the 1970s. And so did the use of solar power generally. As of 2007, along with a substantial number of homes with solar thermal hot-water systems installed, fully half a million homes in Southern California were receiving direct solar power, either from solar electricity plants or from rooftop photovoltaic panels.

These are changing times. Students, workers, consumers, businesspeople, and government officials are increasingly aware that humans have been placing terrific stresses on the environment, as global warming, water shortages, mass extinctions of plant and animal species, and the dwindling supply of fossil fuels attest.

In times of trouble, uncertainty, and even crisis, as entrepreneurs will tell you, there are opportunities. And, as the three examples above suggest, there are reasons to be optimistic, for a rising generation of students, workers, consumers, businesspeople, and government officials—people, in short, from every walk of life and from every corner of the world—is more and more committed to the idea that we can bring the most important power of all to bear on the problem of where our power comes from—that being brainpower.

Green transportation is just one of the sectors where jobs can be found. *Photo: Hydrogen fuel cell bus in Amsterdam; courtesy of Shell Hydrogen.*

Look at it this way: Lance Armstrong, the great bicycle racer, burns as much energy as a handheld hair dryer. So does a cheetah running at full clip. Our brains use about as much as a refrigerator bulb, whether we're thinking hard or barely sentient. It doesn't tax us too much to think, hard and long, about what we can do to lessen our footprints on the land. We have a big problem at the outset: the world's people use nearly 80 million barrels of oil a day, and there is much work to do to wean ourselves from our addiction to irreplaceable fossil fuels.

This book is about the many ways in which we can do that by working with "soft," renewable forms of energy: power derived from the wind, the sun, the sea, the earth, power that causes little or no damage and that can constantly be replaced. Getting there requires a lot of brainpower—and some elbow grease, too.

Jobs in many of those renewable forms of energy were once few and far between. These days, though, companies are going lean, clean, and green. In every profession, in every walk of life, individual consumers, institutions, and corporations are looking to do things better and smarter in the work of producing and using energy and in lessening our dependence on fuels that will one day run out—perhaps sooner than we think.

As high school and college graduates will quickly discover, all other things being equal, there's no better time to learn environmentally responsible habits and put them to work in the marketplace. "Green jobs" are plentiful, and demand for green things is on the rise. New fields are emerging as a result: a few years ago, there was no such thing as someone who brokered carbon trades, for instance, and very few builders and buyers who used terms such as "green buildings."

The fastest-growing professions within the green-jobs realm, according to Kevin Doyle, president of the Boston-based consulting company Green Economy, are environmental engineers, hydrologists, environmental-health scientists, and urban and regional planners. Employers with such positions to fill look for students who major in engineering, mathematics, earth sciences, environmental studies, public policy—but also who come from liberal arts or science backgrounds generally, as long as they can think through a problem and communicate solutions to it.

One gauge of this is the composition of the student body in Yale

Honda's Fuel Cell Scooter equipped with Honda FC Stack. *Photo: Honda*

Google Goes Green in a Big Way
Google wanted to find a way to reduce energy costs at its Mountain View "Googleplex" as well as make a statement in support of clean energy. E1 Solutions engineered a way to mount 10,330 Sharp solar modules on every available rooftop plus the carports. By installing the largest solar power system ever installed at a single corporate campus, the 1.6 megawatt-system will save Google more than $393,000 annually in energy costs, and the system will pay for itself in approximately 7.5 years. And CO_2 emissions are reduced by 3.6 million pounds per year!
Source: E1 Solutions, www.eispv.com

University's graduate program in environmental studies. Almost all of the students had worked for a few years in a green job before returning for graduate work. They came from these undergraduate backgrounds: 25 percent majored in environmental and earth sciences, 23 percent in natural sciences, 22 percent in social sciences and business, 18 percent in humanities, 7 percent in physical sciences and engineering, and 5 percent in forestry science and management. The door to a green future, it would seem, is wide open to students from many different fields.

The green-job phenomenon is also an important aspect of the trades these days, and skilled workers are needed to build, install, operate, and maintain the workings of green energy: rooftop solar panels, geothermal piping, wind turbines, fuel cells, wave rotors, radiant flooring, the list goes on and on. In particular demand are workers capable of erecting buildings that meet the requirements of the U.S. Green Building Council. Even as the market for new-home construction came to a screeching halt at the end of 2007, reports the *Los Angeles Times*, California's residential real-estate market saw a curious twist: solar-powered homes were outselling traditionally electrified new homes in several markets, and developers were stepping up their use of the technology.

Going green involves all of

The residential and commercial solar and wind industries are growing fast and need qualified workers. *Above: 4.1 kW residential PV array; courtesy of Mile Hi Solar (www.milehisolar.com). Right: courtesy of NEG-Micon*

us. The world is in trouble. Or, perhaps better put, we're in trouble in the world. Getting us out of our current fix is going to take a lot of thought, a lot of work—and a lot of energy.

Careers in Renewable Energy

In this book, we'll look closely at dozens of careers within the broad renewable-energy field. Where possible, we'll use firm figures and talk dollars and cents. For instance, according to the *Environmental Business Journal*, the green industry in the United States in 2005 was about $265 billion, employing 1.6 million people. Green businesses have been growing at a rate of about 5 percent annually since then, which suggests that something like 85,000 new jobs are opening up each year. Some of these jobs are new. Others are new wrinkles on old fields such as the law, architecture, or business administration. Some are quite old: we need farmers to produce the makings of biofuel, for instance, and smooth-talking politicians to convince their colleagues to part with research money and provide grants and tax incentives so that homeowners and businesses can retrofit to take advantage of the energy-efficient, clean technologies of today and tomorrow.

> ### Life-Cycle Engineering
>
> A quiet revolution is underway in engineering classrooms in Canadian universities, prompted by a climate change crisis that has unleased a torrent of new green technologies. It's called Life-Cycle Engineering. And, simple put, it means engineers will design products that can be manufactured while leaving a minimal environmental footprint. LCE involves estimating the environmental costs and social benefits at every stage of the life cycle starting with raw material extraction, through product processing, transport, distribution and finally disposal. More importantly, this "cradle-to-grave" approach to engineering allows the engineer to sift through the hype and critically analyze the true impact of so-called green technologies – their promise and their pitfalls.
>
> — SUMITRA RAJAGOPALAN, *THESTAR.COM*

And, as always, we need inventors, entrepreneurs, and visionaries to bring us new technologies and spread the word about them—the same people who, once upon a time, brought us electricity. We hope that some of the information you'll find in the pages ahead will inspire those kinds of people to get going on the big solutions we need to our big problems—and the big opportunities they occasion.

Knowledge Is Power

We also focus here on some of the places that offer training for entering the world of renewable energy: trade schools, community colleges, four-year colleges and universities, and graduate programs, along with professional organizations that sponsor workshops, conferences, and other continuing-education opportunities.

Because so many aspects of the green-energy field are new, there are not always abundant programs to match them. At MIT, for instance, that world-renowned school, there is no major as such in "renewable energy," though the interested student will find plenty of fields, from engineering to physics to business, that deeply concern themselves with renewable-energy matters. With guidance, a student could even design his or her own program with some of the courses scattered about in several departments, such as:

Solar energy at Ell Student Center, Northeastern University – Boston, Massachusetts. *Photo: Ascension Technology, Inc.*

◆ Alternate Energy Sources

◆ Ecology I: The Earth System

◆ Ecology II: Engineering for Sustainability

◆ Fundamentals of Advanced Energy Conversion

◆ Material Science

◆ Mechanical Engineering

◆ Multiscale Analysis of Advanced Energy Processes

◆ Ocean Wave Interaction with Ships and Offshore Energy Systems

◆ Sustainable Energy

Not everyone can go to MIT, of course. Not everyone wants to, and not everyone needs to. Those who live near Flagstaff, Arizona, can get a fine start on making a career in renewables at Coconino Community College, which offers a certificate program in alternative energy technology in which students, the catalog tells us, "are introduced to design issues associated with home construction, community development and passive solar design." That program is made up of required courses that are as rigorous as MIT's, at first glance:

◆ Business English
◆ Building the Human Environment
◆ Construction Materials and Equipment Safety
◆ Building Construction Methods I
◆ Solar Home Design
◆ Intermediate Algebra

Where To Study Renewable Energy

Scott Sklar is president of The Stella Group, Ltd, which facilitates clean distributed energy, including advanced batteries and controls, energy efficiency, heat engines, minigeneration, microhydropower, modular biomass, photovoltaics, small wind and solar thermal. He was previously executive director of two national trade associations, the Solar Energy Industries Association and the National BioEnergy Industries Association. He also cofounded the U.S. Export Council for Renewable Energy. His book *Consumer Guide to Solar Energy* is in its third printing.

We asked him to name the ten schools he would recommend to a student interested in pursuing a career in renewable energy, and these are his picks (see appendix for contact info):

◆ Arizona State University
◆ Bradley University
◆ Colorado State University
◆ George Washington University
◆ Massachusetts Institute of Technology (MIT)
◆ North Carolina State University
◆ Syracuse University
◆ University of California (Merced)
◆ University of Central Florida
◆ University of Nevada

Add another year's study to that made up of the following courses, and you'll earn an advanced certificate:

◆ Introduction to Computer Information Systems

◆ Technical Drafting & CAD Fundamentals

◆ Micro Economics Principles

◆ Blueprint Reading

◆ Building Construction Methods II

◆ Photovoltaics and Wind Power

◆ Innovative and Alternative Building Techniques

◆ Technical Problem Solving

That's quite a full plate. Any program of study is going to be. The payoffs are immediate, though, not just in the sense of preparing a student to enter the renewable-energy field, but also in rewarding that student for his or her labors and preparation. A solar-panel installer with a high-school diploma will be able to earn a reasonable hourly wage, but without further training that wage is liable to remain steady in years to come. With further training, such as this national certification workshop in solar hot water heating, which calls for a high-school diploma and plumbing skills, the pay is likely to go up. Here's the ETM Solar Works program (see *www.etmsolar.com* for more information):

Installing a solar hot water system on a home's roof.
Photo: Steven C. Spencer, FSEC

SOLAR HOT WATER COURSE OUTLINE
DAY 1
Introduction - Solar Power Overview (1 hr) *Different types of solar energy systems; Various types of solar thermal systems descriptions.*

Sunshine Basics (2 hrs) *Path of sun; Finding solar*

noon; *What is full sun; How much full sun is received?; Best location for solar collectors; Shading issues, using Solar Pathfinder; Effects of off-south installation; Collector tilt angles; Understanding thermal units.*

System Sizing (1.5 hrs) *Sizing a solar hot water system for certain hot water usage; Understanding collector specifications and listing; NREL data bases for estimating system performance; Work on computers—learning how to make calculations with spreadsheet.*

System Design (2.5 hrs) *Closed-loop, Thermomax, Thermodynamics LTD, and Solahart systems; System components; Collector mounting schemes; Heat exchangers and storage tanks sized for number of collectors; Pump sizing; Function of various valves and gauges; Antifreeze; Controllers; Types of pipe to use; Pipe routing.*

DAY 2

Building Codes and Zoning Laws (1 hr) *Applying for building permits; Working with local code inspectors; Working with architecture and historical boards.*

Close-loop Fill Assembly (2 hrs)
Hands-on fill assembly; the art of soldering will be addressed as needed; A couple of types of heat exchangers and collectors will be on hand to study; Demos: Thermomax tube, innards of flat-plate fin tube collector, Sunmaster tube, Hotrod heat exchanger, Spirex heat exchanger.

Plumbing and Heating Code (1 hr) *We will recruit union plumber/trainer to discuss code issues; A few electrical code issues will also be covered.*

Mechanical Considerations for Mounting Panels (1 hr) *Wind loading calculation; Evaluating integrity of*

Flat plate collectors of a laundromat's solar thermal system. *Photo: Steven C. Spencer, FSEC*

Energy Consumption per Person
In 1949, energy use per person was 215 million Btu. In 2006, it was 334 million Btu.

SOURCE: ENERGY INFORMATION ADMINISTRATION

roof due to weight of panels and wind loading; Mounting of PV panels on roof rack mounts.

Safety Issues (1 hr) *Working on roofs; OSHA Requirements/ Guidelines; Ladder, scaffold & other safety issues.*

Selling Solar Power (1 hr) *Good business practices; Solar Hot Water Economics; Sales techniques; Aesthetics issues.*

DAY 3
Build Solar Hot Water System

Review *Typical flow of process to put up a residential system from beginning to end; Questions and answers.*

Exam *Proctored, open-book, 2-hour limit.*

Evacuated solar tubes on the roof of this home will heat their domestic water.
Photo: Rex Ewing

Consumers can cut their energy bills by up to 50 percent by investing in solar water-heating technologies.

SOURCE: ENVIRONMENT CALIFORNIA RESEARCH & POLICY CENTER

Whew! That's quite a plateful, too. But you can bet that anyone who follows such a course of study and gains certification will greatly enhance his or her earning potential.

Just so, those with college training have greater earning power, generally speaking, than those without it. According to the University of Wisconsin Engineering Career Service (*https://ecs.engr.wisc.edu*), for instance, in 2007, a man or woman holding a bachelor's degree in civil and environmental engineering and entering the profession earned a median of $48,018 at the outset; a degree in electrical engineering commanded $55,400, and one in mechanical engineering brought slightly less, $54,000. Add a master's degree to the qualifications, and those figures climbed to $56,500, $79,500, and $64,500 respectively. Add a Ph.D., and the entry-level pay for an electrical engineer was $98,000, and that of a mechanical engineer $76,500. The school did not place a civil and environmental engineer in 2007, so no figure is available—but we can assume it would show a similar spread.

Knowledge is power. And, in the renewable-energy field, knowledge is money, too.

Resources

Each chapter in this book includes resources for further study, listing professional organizations, think tanks, study groups, web sites,

LEED Gold: The Solaire 27-Story Residential Tower

Consider the variety of skilled workers needed to design and construct this Gold-certified LEED building in New York:

The Solaire—a 27-story residential tower with 293 units—cut its energy demand by 35 percent using automatic dimming fluorescent lights, high windows, daylighting and other strategies; west-facing photovoltaic panels supply 5 percent of the building's energy needs. Ninety-three percent of the construction waste for the project was recycled and about 60 percent of the building materials were made from recycled content. To maintain superior air quality, the building fea-

tures filtered fresh air, operable windows and controlled humidity.

Its residents have access to public transportation, on-demand hybrid rental cars, bicycle parking and electric vehicle charging. Gardens of native shrubs, perennials and bamboo cover 75 percent of the roof, helping to lower heating and cooling loads and increase tenant satisfaction. To help reduce potable water demand by 50 percent overall, the building uses recycled wastewater for its cooling tower, low-flow toilets and for irrigating landscaping.

The Solaire is an Albanese Organization development. For more information: *www.albaneseorg.com* or *www.thesolaire.com*. *Photo courtesy of Albanese Organization.*

and publications. You'll find specialized listings there. Here we'll simply list some good sources of information about renewable energy generally, all good places to start studying this vast and fascinating field. ❖

General Resources

ORGANIZATIONS & WEB SITES

Business for Social Responsibility

BSR, founded in 1992, guides companies of all sizes and sectors in sustainability issues. A leading global resource for the business community, BSR equips its member companies with the expertise to design and implement successful, socially responsible business policies, practices, and processes, including matters concerning energy.
www.bsr.org

Energy Information Administration

The EIA offers a wealth of information on all aspects of renewable energy, including reports and resources for professionals and students alike.
www.eia.doe.gov

Environmental Career Center

ECC has been helping people work for the environment since 1980. In addition to their web site, ECC publishes *Green Careers Journal*.
www.environmentalcareer.com

Environmental Career Opportunities

A web site with job listings in various environmental categories, including renewable energy.
www.ecojobs.com

Environmental Careers Organization

ECO works to enhance the development of environmental careers through internships, career advice, career products, and research and consulting.
www.eco.org

Green Energy Jobs

This international site offers a global platform for the latest job vacancies in the renewable energy sector, which includes wind, wave and tidal energy, bioenergy, solar, green building, and much more.
www.greenenergyjobs.com

Greenjobs

"We created Greenjobs to ease the path of talent into the renewable energy industries, to ensure that their growth is never stifled through lack of people," says Peter Beadle, the organization's president. The site is one of the premier online job

boards, and a source of news about people in the renewable-energy field.
www.greenjobs.com

Renewable Energy Access
For nearly 10 years, this comprehensive web site has provided access to daily renewable energy news, products and techology overviews, calendar of events, and an extensive listing of RE job opportunities.
www.renewableenergyaccess.com

SustainableBusiness.com
An excellent web site that provides global news and networking services to help green business grow. They offer visitors a unique lens on the field as a whole, covering all sectors that impact sustainabilty: renewable energy/ efficiency, green building, green investing, and organics. Check out the Dream Green Jobs section for a variety of positions.
sustainablebusiness.com

PUBLICATIONS

Godfrey Boyle, ***Renewable Energy***, 2nd edition (Oxford University Press, 2004)
The prospect of producing clean, sustainable power in substantial quantities from renewable energy sources is now arousing interest worldwide. This textbook provides a comprehensive overview of the principal renewable energy sources, including: solar-thermal, photovoltaics, bioenergy, hydro and tidal, wind, and geothermal. With abundant tables, illustrations, and case studies, it provides an interdisciplinary approach to the subject.

Oil for Transportation
In 2006, transportation got 96 percent of its energy from petroleum.

Coal for Electricity
In 2006, 99 percent of coal's energy went to electric power.

SOURCE: ENERGY INFORMATION ADMINISTRATION

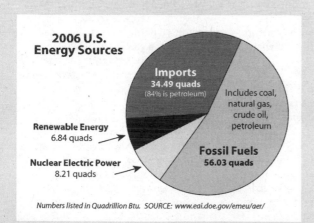

2006 U.S. Energy Sources

Imports
34.49 quads
(84% is petroleum)

Includes coal, natural gas, crude oil, petroleum

Renewable Energy
6.84 quads

Nuclear Electric Power
8.21 quads

Fossil Fuels
56.03 quads

Numbers listed in Quadrillion Btu. SOURCE: www.eai.doe.gov/emeu/aer/

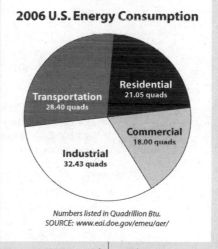

2006 U.S. Energy Consumption

Transportation
28.40 quads

Residential
21.05 quads

Commercial
18.00 quads

Industrial
32.43 quads

Numbers listed in Quadrillion Btu.
SOURCE: www.eai.doe.gov/emeu/aer/

Daniel D. Chiras, **The Homeowner's Guide to Renewable Energy: Achieving Energy Independence Through Solar, Wind, Biomass and Hydropower** (New Society Publishers, 2006)
Chiras discusses energy conservation and renewable energy options, including solar hot water, cooking, and water purification; passive and active solar retrofits; wood heat; wind-generated electricity; electricity from microhydropower sources; and emerging technologies such as hydrogen, fuel cells, methane digesters, and biodiesel.

Aldo V. da Rosa, **Fundamentals of Renewable Energy Processes** (Academic Press, 2005)
This book offers the technical detail necessary to understand the engineering principles that govern renewable energy application at many different levels. Focused on the fundamental mechanisms and processes that underpin energy management, it provides students with the foundation for all energy-process courses.

Rex A. Ewing, **Power With Nature 2nd edition: Alternature Energy Solutions for Homeowners** (PixyJack Press, 2006)
Written for homeowners, this book introduces you to solar, wind and microhydro energy and how it can be used in off-grid and grid-tied homes. Includes chapters on pumping water and heating the home, plus comprehensive appendices.

Phillip F. Schewe, **The Grid: A Journey Through the Heart of Our Electrified World** (National Academy of Sciences, 2007)
"Taken in its entirety, the [electrical] grid is a machine, the most complex machine ever made. The National Academy of Engineering calls it the greatest engineering achievement of the 20th century. It represents the largest industrial investment in history." So writes Schewe, a physicist, at the outset of this fascinating tour of our country's massive, endlessly complicated electrical system. Essential reading for all students of energy, at whatever level.

Jefferson W. Tester, Elisabeth M. Drake, Michael J. Driscoll, Michael W. Golay, and William A. Peters, **Sustainable Energy: Choosing Among Options** (MIT Press, 2005)
This textbook is designed for advanced undergraduate and graduate students as well as others who have an interest in exploring energy resource options and technologies with a view toward achieving sustainability. Chapters include resource estimation, environmental effects, and economic evaluations; the main energy sources of today and tomorrow, including fossil fuels, nuclear power, biomass, geothermal energy, hydropower, wind energy, and solar energy; energy storage, transmission, and distribution; the electric power sector; transportation; industrial energy usage; commercial and residential buildings; and synergistic complex systems.

Solar Energy

1 Solar Energy

As of 2007, solar energy provided less than one percent of the power used in the North American grid. As of 2027, it is estimated, as much as a quarter of America's energy mix may come from solar sources. There are obstacles to be overcome, some technological, some institutional—and those obstacles mean opportunity at many levels, from the hunger of an audience looking for better and more efficient ways to capture the sun's energy to the

Photo courtesy of DC Power (www.dcpower-systems.com)

need for skilled workers in installing, maintaining, and operating solar-energy systems. These systems may be massive, such as the great solar plant at Kramer Junction, California, or small, such as a simple set of rooftop-mounted solar panels capable of providing some or all of a single home's electrical power needs. Whatever the case, these systems require well-trained, smart people to develop and improve them and to keep them running efficiently.

Solar energy works by converting energy from the sun, in the form of light, into electricity. This is accomplished by several means. One is to use photovoltaic (PV) cells made up of semiconductors.

Rooftop solar panels, for example, are typically made of multiple cells encased in weather-resistant housing, and these cells convert sunlight into direct current. (Other kinds of panels are used to heat water directly; these solar thermal systems do not generate electricity but are used to heat homes and pools.) Another is to channel sunlight onto fluid to cause it to boil and thereby spin a turbine; at the Kramer Junction plant, for instance, a field of highly polished mirrors concentrates reflected light onto a synthetic oil that passes through water to produce clean-burning steam that in turn drives a turbine, powering thousands of nearby houses. Still another is to use engines that are directly powered by solar heat, such as the Stirling engine, in which gas heated by the sun pushes a piston that in turn powers a generator; as the gas leaves the heat source it cools, and eventually it returns through the system to go back into the cycle anew.

The future will doubtless bring new means of using the sun's energy to make power that is directly useful to us. It is up to physicists, material scientists, engineers, and inspired inventors from inside and outside academia to come up with those new means. It is up to technicians and workers of many kinds to get the results of that engineering online and available to us, and to keep it running. Many kinds of thought and experience are needed in the field—and many kinds of careers in solar energy await people from just about every background imaginable.

A growing population requires homes, shops, and workplaces, and this in turn opens many prospects

Above: A Sandia engineer manually operates the solar concentrator at Ft. Huachuca, Arizona. Right: A dish Stirling system at a Golden, Colorado test site. *Photos: Sandia National Laboratory*

for those interested in careers in installing and maintaining energy systems. Industry estimates indicate continued strong growth in solar energy jobs worldwide, noting that most of those new jobs will come in marketing and installing solar photovoltaic and thermal systems for homes, businesses, schools and hospitals, and other facilities. Germany represents a case in point: in 1998, the German solar industry employed about 1,500 people, while in 2007 the number was nearly 25,000, with an anticipated 100,000 jobs created by the beginning of the next decade. This requires both political will and money, of course, which Germany has backed up by an ambitious program of federal loans and other economic incentives. Other European countries are projecting a similar growth rate. "By the time the generation born today reaches adulthood in 2020," the European Photovoltaic Industry Association declares, "solar energy could easily provide energy to over a billion people globally and provide 2.3 million full-time jobs."

Just so, the market for solar energy in the United States has grown markedly. In 1999, there were only 18 megawatts of photovoltaic (PV) installations across the country (a megawatt is a million watts); that number had increased to 66 megawatts by 2003, about one-tenth of the output

Solar Home Facts

A 2.2-kilowatt home solar system that consists of twelve 180-watt photovoltaic panels requires about 225 square feet of installation area, easily available on most roofs. This system will generate approximately 2,800 kilowatt hours per year in a sunny climate, about half that in a cloudy climate. The consumer doesn't have to bear this cost alone, thanks to local and state government incentives such as grants and tax breaks.

A very large residential solar array. *Photo courtesy of Sharp (http://solar.sharpusa.com/solar)*

of a single solar-power plant in New Mexico slated to go online in 2011. California plans to add about 300 megawatts annually from 2007 onward, and by 2010, it is estimated, the cumulative installed capacity in solar power will be 1.6 gigawatts (a gigawatt is a billion watts) in the United States and 9.5 gigawatts worldwide.

Careers in Solar Energy

Within the solar-energy field, there are white-collar, blue-collar, and what might be called green-collar jobs, altogether making a mix of administrative, scientific, business, and technical skills. Where you land on the ladder largely depends on your interests and abilities, as well as on the course of education and specialized training you follow.

Administrative, Financial, and Nontechnical

Most jobs in the white-collar realm require, at a minimum, a degree from a four-year college. Many workers at the higher levels of utility administration, for example, hold degrees in business or economics. Chief executive and business-development officers usually have a solid education combined with broad professional experience of at least ten years' time; finance directors, accountants, and corporate strategists have excellent number-crunching skills. In all white-collar levels of the energy business, good mathematical and communication skills are considered to be of great importance, since an executive has to have a grasp on business fundamentals and, ideally, should be able to communicate the vision of his or her company to the outside world and constituents within the organization.

Other nontechnical jobs are as numerous and varied as technical ones within the field. Companies large and small require marketing,

A power plant producing a continuous megawatt (MW), or one millions watts, of electricity will provide electricity to power approximately 800 conventional homes, or over 1,500 energy-efficient homes.

public relations, and human-resources services, and many careers in the renewable-energy market will involve sales, or a combination of sales and technical skills. One Southern California employer recently ran an ad for an inside sales position, for example, that called for "strong leadership skills, a positive, can-do attitude, 2 to 5 years of experience in business, world-class communications and customer-service skills, the ability to work long hours, and strong math and computer skills." Added to this was the strong preference that the successful applicant hold a four-year college degree. The responsibilities of the job, the employer specified, were these:

- Learn about the solar industry, current technology and the rebate programs in order to effectively respond to general inquiries from potential customers.
- Manage incoming new leads that arrive by email/phone; Manage general email communication.
- Qualify new leads by completing customer questionnaire.
- Add customers into customer database.
- Log all communications with customers in customer database.
- Assist in special projects as needed, perform research and maintain records.
- Coordinate mass mailing, conferences, and marketing efforts as required by program departments.
- Serve as backup office manager.
- Set appointments for site evaluations.

An NREL engineer tests the output of PV panels.
Photo: Holly Thomas, NREL

In a similar spirit, the Department of Energy's National Renewable Energy Laboratory (NREL) recently announced a job opening for a technical writer and editor, another nontechnical job that, of course, has many specialized requirements of its own. At a minimum, the posting specified:

> Bachelor's degree in communications, journalism, or English or equivalent experience, plus five years of relevant work experience in a scientific or technical information environment. The selected candidate must have a good understanding of communications theory and its application and must be able to provide highly detailed editing of technical and outreach print publications and web sites. Selected candidate will conduct research and write technical and outreach material for a variety of audiences, including the general public, interested stakeholders, research partners, and policy makers. Demonstrate project management and business development skills and work on multiple communications projects simultaneously. Personal computer skills such as word-processing, Internet browser, and e-mail applications are required.

Research

Solar-energy research positions call on academic and business skills of many kinds. A look at the NREL home page (*www.nrel.gov*) quickly shows that a solid grounding in mathematics, physics, chemistry, and other sciences is essential to most such positions within the field.

Consider what sort of background might be required, for instance, to enter the following program, as described on NREL's site:

> The U.S. Department of Energy (DOE) researches and develops a clean, large-scale solar thermal technology

Top: An NREL researcher uses an infrared microscope to examine novel PV materials. *Photo: Mike Linenberger, NREL* Bottom: Research is being conducted on thin film solar cells. *Photo: Institute of Energy Conversion*

known as concentrating solar power (CSP). This research and development (R&D) focuses on three types of concentrating solar technologies: trough systems, dish/engine systems, and power towers. These solar technologies are used in CSP plants that use different kinds of mirror configurations to convert the sun's energy into high-temperature heat. The heat energy is then used to generate electricity in a steam generator.

Concentrating solar power's relatively low cost and ability to deliver power during periods of peak demand—when and where we need it—mean that it can be a major contributor to the nation's future needs for distributed sources of energy.

DOE's Solar Energy Technologies Program pursues concentrating solar power R&D to provide clean, reliable, affordable solar thermal electricity for the nation. The National Renewable Energy Laboratory (NREL) and Sandia National Laboratories work together as SunLab, a partnership developed by DOE to support R&D within the Concentrating Solar Program.

Installation of a mirror panel on a concentrating solar power system in New Mexico. *Photo: C.E. Andraka, Sandia National Lab*

One indication might be NREL's requirements for an entry-level engineering position:

Bachelor's degree in engineering or closely related field, or equivalent relevant experience, plus a demonstrated understanding of relevant R&D. Fundamental knowledge of engineering practices, procedures, and concepts. Basic engineering abilities in practices and techniques; applies basic technical skills and methods to help analyze and solve problems. Demonstrated adequate interpersonal and communication (oral and written) skills. Must be computer literate including programming, interfacing, and software development and/or utilization.

At a far more advanced level leading to a senior research position, NREL advertised for a postdoctoral fellow in photovoltaics, the inner workings of solar panels:

> A recent Ph.D. (less than three years), in Chemistry, Chemical Engineering, or Physics. Candidates must have: a strong background in ink-based materials, solution patterning, liquid processing and thin film deposition; strong scientific writing skills demonstrated by experience in writing quality scientific papers; strong communications skills; and, the ability to work collaboratively as part of a research team.

Manufacturing process of solar cells. *Photo: Shell Solar Industries*

Similarly, Heliovolt, Inc., a new company in Austin, Texas, specializing in the manufacture of thin, light solar installations, recently listed a job in thin-film circuit manufacture. The job calls for a master's or doctoral degree in physics or applied math, as well as "experience designing and modeling thin film circuits, strong mathematical background, experience using modeling and simulation tools (SPICE, Mathematica, etc.), and a general knowledge of thin film fabrication processes."

Such skills are hard to attain, naturally enough. Without such skilled workers and thinkers, though, the renewable-energy field cannot advance, and those who are best prepared will be the most successful.

By every projection, it's a seller's market for technicians and other crafts workers well trained in installing and maintaining solar-energy systems. Demand is expected to exceed supply considerably over at least the next decade, and though the entry-level qualifications are rigorous, the rewards are substantial.

Here, for example, is an excerpt from a help-wanted listing posted by a solar-energy contractor in southern California, calling for a solar systems installer. The desired qualifications, the employer says, are these:

◆ Strong work ethic, self motivation, organization and a can-do attitude.

◆ Experience in mechanical installations, general construction, and ladder work.

◆ Basic understanding/experience with electrical wiring of AC and DC systems (preferred).

◆ 1–2 years of construction background is preferred.

◆ Experience with all types of hand-held and power tools.

◆ Experience with small machinery (trenchers, bobcats, forklifts) preferred.

◆ Experience working with all types of building materials— various roof types, stucco, wood, concrete, Uni-strut, roofing etc. preferred.

◆ Team player who listens, learns, and actively communicates.

Installing a ground-mounted solar array. *Photo: Sierra Solar (www.sierrasolar.com)*

◆ Visual thinker good at problem solving and implementing ideas.

◆ Knowledge of the renewable-energy marketplace, technology, and industry.

◆ Desire to learn and master all aspects of installing solar photovoltaic systems.

◆ High-school diploma, two-year degree in technology/industrial arts preferred.

◆ Employment with a well-known contractor preferred.

Connecting the large solar arrays at a rice drying facility in California. *Photo: DC Power (www.dcpower-systems.com)*

That's quite a range of skills, all requiring work, discipline, and study to attain. The job duties are just as varied; the employer asks that the employee perform project installations including on-site design implementation, assembling mounting hardware, mounting equipment, wiring solar systems, and documenting the work done in words and photographs, as well as cleaning up the site, servicing tools and vehicles, and talking about and selling the employer's products "to interested persons as required." That's quite a full plate, and we can only hope that the employer is willing to pay well for the right person to pick it up.

In its listing, the same employer indicates that NABCEP solar PV installer certification will go a long way in meeting all its qualifications. This certification comes from the North American Board of Certified Energy Practitioners, a trade association organized, in part, to oversee standards within the field. To acquire this NABCEP certification, a candidate must be at least 18 years of age; meet prerequisites of related experience and/or education; complete an application form documenting requirements; sign a code of ethics;

pay applicable fees; and pass a written examination. The NABCEP writes that there are seven paths to meeting the experience and/or education requirement:

- ◆ Four (4) years of experience installing PV; OR
- ◆ Two (2) years of experience installing PV systems in addition to completion of a board-recognized training program; OR
- ◆ Be an existing licensed contractor in good standing in solar or electrical-construction related areas with one (1) year of experience installing PV systems; OR
- ◆ Four (4) years of electrical-construction related experience working for a licensed contractor, including one (1) year of experience installing PV systems; OR
- ◆ Three (3) years experience in a U.S. Dept. of Labor approved electrical-construction trade apprentice

Women's Solar Workshops

Solar Energy International's "Women and Renewable Energy Program" provides an opportunity to break the chain of gender bias. Training programs teach renewable energy technologies to equip women with the necessary skills to participate in the design, installation and maintenance of systems. "We've had NASA engineers, licensed electricians, schoolteachers, and homemakers attend our women's PV Design & Installation workshops," says Laurie Stone of SEI.

Assembling a PV mounting rack.
Photos: Solar Energy International

program, including one (1) year of experience installing PV systems; OR

- ◆ Two-year electrical-construction related, or electrical engineering technology, or renewable energy technology/technician degree from an educational institution plus one (1) year of experience installing PV systems; OR

- ◆ Four-year construction related or engineering degree from an educational institution, including one (1) year experience installing PV systems.

Delivering Solar Electricity to the Rural-Poor

Neville Williams founded the nonprofit Solar Electric Light Fund (SELF) in 1990. It promoted solar power for a decade by setting up pilot solar rural-electrification programs in eleven countries. In 1997 he launched the commercial Solar Electric Light Company, which has brought solar electricity to over 75,000 families in Asia, chiefly India, Sri Lanka and Vietnam. His experiences and insights can be found in his book *Chasing the Sun: Solar Adventures Around the World* (New Society Publishers, 2005). Here are some excerpted passages:

Ironically, people who never benefited from the age of oil, who never had electricity, are the solar power pioneers who today are using the technology we will all be using in the future. They are building the much-touted "solar economy."

We could deliver an SHS [solar home system] for less than it cost to run an electric line 100 meters from the nearest power main. And this didn't include the cost of transmission or generating the power, just the

hookup and house wiring. The sun is, indeed, a better distributor of power than copper wires, especially when houses are widely dispersed, as they are in rural Sri Lanka and in most developing countries. Sri Lanka became our solar learning lab, and today it leads the world in solar installations. One day it may become the world's first solar-powered island.

Over the years, I personally listened to hundreds of personal testimonials from peasants as they described how electricity changed their lives, how they no longer needed to use kerosene, how much better their children could study at night, and how much they enjoyed seeing television for the first time. One I recall was an old man who said, having seen Jiang Zemin on TV, "I have never before been able to see the emperor!" Another elderly peasant, shedding tears, said, "I have long heard that city folks do not need oil to generate light, but in all my 70 years, this is the first time to actually see such a phenomenon with my own eyes. What a beautiful thing!"

Other trades within the solar-energy field involve skill in crafts such as plumbing, electricity, and carpentry. Jobs include the design and maintenance of solar hot-water systems, using the sun to heat domestic hot water as well as buildings themselves; the construction of buildings and communities that use renewable energy; and the retrofitting of older homes and businesses to accommodate new technologies.

Knowledge Is Power

The job market in the renewable-energy sector is expected to be strong well into the 21st century, especially for those who enter it with training and certification from an accredited technical school or with formal apprenticeship training. Economists warn that technicians who specialize in installation work may suffer periods of unemployment, since new construction is susceptible to the whims of the market, but that maintenance and repair work will almost always offset times when the economy goes soft. Newer systems tend to be more complex than their predecessors, which is one reason that employers are likely to prefer technical-school graduates or students who have taken appropriate courses at the junior and community-college level.

Another home gets a grid-tied solar array. *Photo: Standard Solar, Inc. (www. standardsolar.com)*

Fortunately, many secondary and postsecondary technical and trade schools, junior and community colleges, and programs within the armed forces offer training in aspects of solar-power systems, including heating, air-conditioning, and refrigeration. In them, students also acquire at least a basic understanding of electronics. On the trades level, formal apprenticeship programs are also offered through organizations such as the Associated Builders and

Contractors, the National Association of Home Builders, and the International Brotherhood of Electrical Workers. These programs normally last from three to five years and combine on-the-job training with classroom instruction. Under most circumstances a candidate for them must have a high school diploma or the equivalent and have solid math and reading skills, for in those trades the ability to measure, specify, and communicate clearly are of the utmost importance.

Specialized undergraduate programs in alternative energy, including solar, are few now, but they are likely to become much more common in the near future. One interesting exception is the Photovoltaics and Solar Energy undergraduate program at the University of New South Wales in sun-drenched Australia, which offers a five-area approach that it describes as follows:

1. Device and system research and development.

2. Manufacturing, quality control and reliability.

3. PV system design (computer based), modeling, integration, analysis, implementation, fault diagnosis and monitoring.

4. Policy, financing, marketing, management, consulting, training and education.

5. Using the full range of renewable energy technologies including alternate energy technologies (such as wind, biomass and solar thermal), solar architecture, energy efficient building design, and sustainable energy.

The photovoltaic engineering program includes training in technology development, manufacturing, quality control, life-cycle analysis, system design, diagnosis and maintenance, and policy development, all critically important fields.

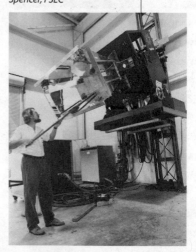

The Florida Solar Energy Center's solar simulator (which is getting a new light bulb) tests solar thermal panels by simulating natural daylight and wind conditions. *Photo: Steven C. Spencer, FSEC*

It weds practical and theoretical knowledge and requires industrial training as well as classroom work.

For scientific and theoretical work, undergraduate and, usually, graduate training is expected and even required. At the top end of the schools in the energy field is the Massachusetts Institute of Technology, which counsels that students prepare themselves well in high school for the rigorous work that follows. The MIT web site suggests an academic foundation that includes:

- One year of high school physics
- One year of high school chemistry
- One year of high school biology
- Math, through calculus
- Two years of a foreign language
- Four years of English
- Two years of history and/or social sciences

Installing PV panels at a rice drying facility in California. *Photo: DC Power (www. dcpower-systems.com)*

"Overall, you should try to take the most stimulating courses available to you," the site continues. "If your high school doesn't offer courses that challenge you, you may want to explore other options, such as local college extension or summer programs."

If, as an MIT student, you wanted to follow a program in electrical science leading to a specialization in solar-electrical engineering, your coursework would follow a sequence requiring classes and labs in probability and differential equations, as well as electromagnetism, signal processing, and device and circuit design. Graduate specialization might follow several options, depending on the nature of the career you want to follow. The advisors for the University of Colorado graduate program in building-systems engineering put it this way: "Your course of study may be wide or narrow; it will depend on your personal career plans. People who are thinking of a future in research work may take a different approach than those who see themselves working in industry or for a consulting engineering firm."

Like many other graduate programs, Colorado's accommodates students from backgrounds other than engineering, though it recommends undergraduate coursework in calculus, thermodynamics and heat transfer, and building mechanical systems. The program offers concentrations in four areas: energy analysis, HVAC (heating, ventilation and air conditioning) systems, illumination, and solar

Looking down the north side of Decathlete Way, the solar houses of Santa Clara University, Universidad de Puerto Rico, University of Texas at Austin, and University of Maryland are visible. *Photo: David Hicks, Solar Decathlon*

The Solar Decathlon

In October 2007, twenty teams from universities in the United States and abroad competed in a "solar decathlon." Selected by the U.S. Department of Energy, the teams designed, built, and operated attractive and energy-efficient solar-powered homes that were gathered on the National Mall in Washington, D.C., as a "solar village."

In a quest to stretch every last watt of electricity that's generated by the solar panels on their roofs, the students absorb the lesson that energy is a precious commodity.

Team Montreal's home consists of a special structural steel frame that is easy to assemble, as they did in preparration for the Solar Decathlon. *Photo: Kaye Evans-Lutterodt, Solar Decathlon*

They use high-tech materials and design elements in ingenious ways, and along the way, they learn how to raise funds and communicate about team activities, and how to collect supplies and talk to contractors. The Solar Decathletes—tomorrow's engineers, architects, researchers, and homeowners—are sharing with us a new vision for living under the sun.

www.solardecathlon.org

2007 Solar Decathlon Contestants

Carnegie Mellon University

Technische Universität Darmstadt

Cornell University

Texas A&M University

Georgia Institute of Technology

Universidad Politécnica de Madrid

Kansas Project Solar House (Kansas State University and University of Kansas)

Universidad de Puerto Rico

Lawrence Technological University

University of Colorado at Boulder

Massachusetts Institute of Technology

University of Cincinnati

New York Institute of Technology

University of Illinois at Urbana–Champaign

Penn State

University of Maryland

Santa Clara University

University of Missouri–Rolla

Team Montréal (École de Technologie Supérieure, Université de Montréal, McGill University)

University of Texas at Austin

The overall winner of the Solar Decathlon 2007 was Technische Universität Darmstadt of Germany (top), with University of Maryland receiving 2nd place (center), and Santa Clara University 3rd place (bottom).

and renewable energies. The last requires these courses, all listed in the civil engineering department:

◆ Introduction to Solar Utilization

◆ Solar Design

◆ Advanced Solar Design—Photovoltaics

◆ Advanced Passive Solar Design

◆ Energy and Environmental Policy

◆ Analysis and Assessment of Renewable Energy Systems

◆ Building Systems Seminar

The Colorado catalog counsels, "You should also be (or become) competent in computer programming, word processing and spreadsheet usage to solve engineering problems." Other programs across the country have much the same expectations, so if you're contemplating a future on the R&D end of the alternative-energy spectrum, it's time to hit the books! ❖

MEDIAN SALARIES/WAGES ❖ Solar Energy

Solar Thermal Designer	$53,611	– $67,577
Solar Thermal Installer	$38,213	– $48,727
Materials Research Engineer	$59,610	– $108,070
Material Scientist	$50,470	– $90,670
Physicist (PhD)	$80,440	– $119,160
Electrical Engineer	$55,040	– $96,410
Power Engineer	$59,040	– $69,240
Certified Electrician	$21.70 – $35.50 / hour	

SOURCES: WWW.PAYSCALE.COM & ECONOMIC RESEARCH INSTITUTE (DECEMBER 2007)

MEDIAN SALARIES/WAGES ❖ Related Fields

Development Coordinator	$31,260	– $39,810
Marketing Manager	$40,310	– $80,730
Public Relations Manager	$34,290	– $92,364
Senior Sales Engineer	$93,640	– $107,540
Technical Writer	$40,090	– $60,140
Consultant, Environmental Engineering	$34,210	– $81,410
Research Engineer, Environmental Sciences	$38,880	– $101,486

SOURCE: WWW.PAYSCALE.COM (DECEMBER 2007)

Renewable Energy in the Developing World

Here in the United States, solar-electric (that is, photovoltaic) systems typically consist of modules on our roofs, connected to the utility grid to generate some portion of our household electricity. But in many parts of the developing world, solar energy is the only source of electricity for a home or a village, because no grid exists.

In the States, average-sized residential solar-electric systems are between 3 and 5 kilowatts (KW). In the developing world, systems of that size could run an entire village or a large community health center. Average home systems in the developing world are 50 to 75 watts, and "large" systems may be 120 watts—a fraction of the size of a typical residential system in the United States.

A solar hybrid project for a Partners for Health clinic in Rwanda (Solar Electric Light Fund and SunEnergy Power International).

Working with renewable energy in the developing world is exciting and well worth the rewards. If you have the desire to become involved, keep pushing for answers. The right opportunity will surely present itself. The need remains huge, with more than 1.5 billion people who live with absolutely no electricity. Bringing electricity to these people can make dramatic changes in their lives, allowing them to educate themselves, start businesses, and improve their standard of living.

Working to give people in the developing world access to electricity is always a two-way street. The people we help have so much to gain by improving their access to education, health care, additional work opportunities, and much more. But we gain as much or more from the experience.

— Walt Ratterman
Excerpt from *Home Power*, June/July 2007

Walt earned a Master of Science degree in Renewable Energy from Murdoch University in Australia, and has extensive renewable energy design and installation experience internationally, including Nicaragua, the Galapagos Islands, southern Ecuador, Peru, Arunachal Pradesh in India, Burma, Thailand, and Rwanda. He is currently the Chief Project Officer & Director at SunEnergy Power Corporation (*www.sunenergypower.com*)

A solar water pumping project in Pakistan (Solar Energy International and SunEnergy Power International).

Photos: Walt Ratterman

Solar Energy Resources

ORGANIZATIONS & WEB SITES

American Solar Energy Society
Founded in 1954, ASES boasts more than 10,000 members across the nation who are at work creating a sustainable energy economy. Their web site also serves as a meeting place for qualified professionals who are able to design, install, and service solar-energy systems. ASES publishes *Solar Today* magazine, a flagship publication in the field. Job listings can be found at *www.ases.org/jobs.htm*
>2400 Central Avenue, Suite A
>Boulder, CO 80301
>(303) 443-3130
>*www.ases.org*

FindSolar.com
A web site that will connect you to renewable energy professionals; sponsored by ASES, Solar Electric Power Association, and Department of Energy.
>*www.findsolar.com*

Florida Solar Energy Center
FSEC is the largest and most active state-supported renewable energy research institute in the U.S. They are an accredited laboratory for testing and certification of solar technologies. Established in 1975, they also conduct research in building science, solar energy, hydrogen and alternative fuels, fuel cells and other advanced energy technologies.

>1679 Clearlake Road
>Cocoa, FL 32922-5703
>(321) 638-1000
>*www.fsec.ucf.edu*

International Solar Energy Society
ISES has been serving the renewable energy community since 1954. A UN-accredited NGO present in more than 50 countries, the Society supports its members in the advancement of renewable energy technology, implementation and education all over the world.
>*www.ises.org*

Midwest Renewable Energy Association
MREA promotes renewable energy, energy efficiency, and sustainable living through education and demonstration. It offers workshops in renewable energy, including wind systems; photovoltaics; solar thermal systems for domestic hot water and space heating; plus photovoltaic, wind, and solar thermal site assessor training and certification.
>7558 Deer Road
>Custer, WI 54423
>(715) 592-6595
>*www.the-mrea.org*

North American Board of Certified Energy Practitioners (NABCEP)
As its site says, "NABCEP is committed to providing a certification program of quality and integrity for the professionals and consumer/public it is designed to serve. Professionals

who choose to become certified demonstrate their competence in the field and their commitment to upholding high standards of ethical and professional practice."

    10 Hermes Road, Suite 400
    Malta, NY 12020
    (518) 899-8186
    www.nabcep.org

Power from the Sun

This site offers an online, revised and updated, and free version of *Solar Energy Systems Design*, a highly useful textbook first published by W. B. Stine and R. W. Harrigan in 1985. It contains other resources that are useful for learning about solar energy, including problem sets and various calculators.

    www.powerfromthesun.net

Solar Energy International

SEI is a non-profit organization offering accredited On-Line and Hands-on workshops in solar, wind and water power and natural building technologies in 17 locations worldwide. SEI provides the expertise to plan, engineer and implement sustainable development projects. Job openings for SEI alumni can be found online under Resources.

    P.O. Box 715
    Carbondale, CO 81623
    (970) 963-8855
    www.solarenergy.org

Solar Living Institute

Established in 1998, the Solar Living Institute is a nonprofit educational organization whose mission is to promote sustainable living through inspirational environmental education. The Institute provides practical education by example and a range of hands-on workshops on renewable energy, green building, construction methods, and other subjects.

    13771 S. Hwy. 101
    Hopland, CA 95449
    (707) 744-2017
    www.solarliving.org

PUBLICATIONS

Jim Dunlop, ***Photovoltaic Systems*** (American Technical Publishers, 2007) A comprehensive textbook for the design, installation, and evaluation of residential and commercial photovoltaic (PV) systems, both stand-alone and grid-tied systems. The included CD-ROM features interactive resources for independent study and to enhance learning.

Rex A. Ewing and Doug Pratt, ***Got Sun? Go Solar*** (PixyJack, 2005) Meant for homeowners and small-systems designers, this user-friendly handbook of grid-tie solar gives a very good idea of the kinds of skills and training that go into making an energy-efficient building.

Christopher Flavin and Nicholas Lenssen, ***Power Surge: Guide to the Coming Energy Revolution*** (Norton, 1994) An interesting discussion of the technologies discussed throughout this book, Flavin and Lenssen's overview suffers only from being a touch dated. Anyone who aspires to a career in alternative energy, however, ought to give it a careful reading.

Home Power Magazine
Dedicated to renewable energy and sustainable living solutions, this magazine provides extensive product information, homeowner testimonials, buyer advice, and "how-to" instructions.
www.homepower.com

James Kachadorian, ***Passive Solar House: The Complete Guide to Heating and Cooling Your Home*** (Chelsea Green Publishing Company, 2006)
Applicable to diverse regions, climates, budgets, and styles of architecture, this book offers proven techniques for building homes that heat and cool themselves. Includes CSOL passive solar design software.

Richard J. Komp, ***Practical Photovoltaics: Electricity from Solar Cells*** (Aatec Publications, 1995)
A comprehensive guide to the theory and reality of solar electricity, as well as a detailed installation and maintenance manual for home projects.

Rob Ramlow and Benjamin Nusz, ***Solar Water Heating: A Comprehensive Guide to Solar Water and Space Heating Systems*** (New Society Publishers, 2006)
This books presents the basics of solar water heating, including solar water and space heating systems, system components, installation, operation, maintenance, and system sizing and siting.

Hermann Scheer, ***The Solar Economy: Renewable Energy for a Sustainable Global Future*** (Earthscan Publications Ltd, 2004)
A powerful and comprehensive account of how the global economy can and must replace its dependence on fossil fuels with solar and renewable energy—and the enormous and multiple benefits that should follow.

Solar Energy International, ***Photovoltaics Design and Installation Manual: Renewable Energy Education for a Sustainable Future*** (New Society Publishers, 2004)
This sturdy reference work covers the basics of solar electricity, photovoltaic applications and system components, solar site analysis and mounting, component specification, and safety issues, among many other topics.

Neville Williams, ***Chasing the Sun: Solar Adventures Around the World*** (New Society Publishers, 2005)
A fascinating account of the author's twelve-year quest to bring solar power and light to people in the developing world who have no electricity. Iluminating reading to all interested in the environment, development, renewable energy, socially responsible business, and our future at the end of the age of oil.

Wind Energy

2 Wind Energy

Since the early 1990s, wind power has become one of the fastest-growing sources of electricity generation in the United States, while other nations have enthusiastically pursued wind power as well. As of this writing, wind turbines have been put in place in 22 states, both onshore and offshore, with more joining the list each year. California and Texas lead the country, but places such as Ohio and New York are making significant advances in adding wind-generated energy to the overall energy mix.

Photo © Neg-Micon

Wind energy, investors and homeowners alike are finding, is clean, efficient, and inexhaustible. In addition, a single turbine of ordinary size—about 250 feet across, that is—can generate some 2 megawatts at peak output; a solar array of the same size can generate only a fraction of that power and, at present, would cost several times more to put on line. It is for those qualities that states, municipalities, and private utilities are flocking to wind power.

Wind energy is the fastest-growing energy generation technology, expanding by 30 to 40 percent annually.

According to the American Wind Energy Association (AWEA), this high level of investment in wind energy will mean more jobs over the next several years in several areas, such as manufacturing and engineering, environmental and consulting services, and sales and marketing. In addition, research and academic positions are likely to grow in number. The situation, in other words, is very much like that of the solar-energy field (see the preceding chapter), and major players in the energy industry, such as General Electric, Siemens, and Shell Oil, are now heavily involved in research.

What Is a Wind Turbine and How Does It Work?

A wind energy system transforms the kinetic energy of the wind into mechanical or electrical energy that can be harnessed for practical use.

Mechanical energy is most commonly used for pumping water in rural or remote locations—the "farm windmill" still seen in many rural areas of the U.S. is a mechanical wind pumper—but it can also be used for many other purposes (grinding grain, sawing, pushing a sailboat, etc.). Wind electric turbines generate electricity for homes and businesses and for sale to utilities.

There are two basic designs of wind electric turbines: vertical-axis, or "egg-beater" style, and horizontal-axis (propeller-style) machines. Horizontal-axis wind turbines are most common today, constituting nearly all of the "utility-scale" (100 kilowatts and larger) turbines in the global market.

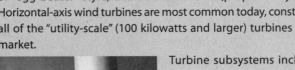

Top: A ranch windmill for pumping water. *Photo: LaVonne Ewing*

Right: Closeup view of a commerical wind turbine that generates electricity. *Photo courtesy of GE (www.gepower.com)*

Turbine subsystems include: a rotor, or blades, which convert the wind's energy into rotational shaft energy; a nacelle (enclosure) containing a drive train, usually including a gearbox and a generator; a tower, to support the rotor and drive train; and electronic equipment such as controls, electrical cables, ground support equipment, and interconnection equipment.

SOURCE: AMERICAN WIND ENERGY ASSOCIATION

All of this suggests a bright future for wind energy, and for those who work in it. For the near future, most new wind-energy jobs will be in manufacturing, installation, and operations: in building, siting, setting up, and maintaining and repairing wind turbines and associated equipment, in other words. There are also opportunities in areas such as marketing, sales, computer modeling, and consulting, and plenty of places to train for careers in the field and enhance one's set of skills.

Careers in Wind Energy

As a comparatively new field, wind energy has not yet solidified into easily defined career tracks. As with solar energy, a technician is often expected to know something of the big picture, with an understanding of trends in the business and promising research. A salesperson, similarly, needs at least a basic understanding of the technologies involved and of the challenges and benefits associated with, say, putting a wind turbine in place. A technical writer will need to know just about everything in a wind-power business, almost as much as a CEO or CFO. And an engineer working in wind power needs to have practical grounding in how a system that looks good on paper can be put up and maintained without bringing the people who do that work to despair. In short, a successful career in wind energy is going to involve a little bit of everything. The person with wide-ranging interests and abilities is going to thrive in this new green-collar world, while the person who can't be bothered to acquire new skills and keep up with developments in the field is likely to hit a dead end very soon.

Small wind energy is also growing at a fast pace, both off-grid and grid-tied installations. *Top photo: Bergey Windpower; Bottom photo: Warren Gretz, NREL.*

How many homes can one megawatt of wind energy supply?
An average U.S. household uses about 10,655 kilowatt-hours (kWh) of electricity each year. One megawatt of wind capacity can generate from 2.4 to more than 3 million kWh annually. A megawatt of wind thus generates as much electricity as 225–300 households use.

U.S. DEPARTMENT OF ENERGY

Administrative, Financial, and Nontechnical

As in most aspects of the energy industry, wind power relies on engineers and technicians, but also on people with business, financial, and communications skills. Holders of bachelor's and master's degrees in business administration, for example, are in demand, while industry insiders also observe that liberal-arts majors with some aptitude for science, business, and math flourish in the industry. Technical writers are also in demand, as well as writers who can explain wind energy and develop campaigns to promote its use.

The American Wind Energy Association (AWEA), for instance, recently advertised for a junior position in advocacy writing, a job that calls for the ability to match industry experts and journalists needing background, as well as writing opinion articles and letters to the editor. The job calls for "strong oral and exceptional written communication skills; detail oriented; ability to proofread; ability to juggle multiple/overlapping projects and meet deadlines; ability to work well with teams; ability to self-motivate; ability to be flexible; ability to establish a clear focus on results." It also requires a bachelor's degree in journalism, communications, English, or marketing, with a preference for applicants with a minimum of three years of communications, marketing, or advocacy work within the wind industry or the nonprofit sector.

Research

At present, according to the AWEA, research opportunities in wind energy center on five broad areas. These are:

 • **Turbine research**: research to improve turbine design (aerodynamics), understanding the nature of wind (inflow and turbulence), and using computer models to design efficient and low-cost turbines (modeling structures and dynamics).

- ◆ **Wind resource assessment**: mapping an area to include specific wind data such as average wind speed and variability.

- ◆ **Forecasting**: using weather models and technologies to predict wind speeds and patterns at various altitudes.

- ◆ **Utility grid integration**: research to improve the ways in which energy produced by wind is introduced into the utility grid.

- ◆ **Energy storage**: developing technology to store wind energy as electricity.

Resource assessment is particularly important and provides many opportunities. It does no good to erect a wind turbine in an area in which the wind does not blow, and trained assessors are essential to the work of determining where to locate wind-generating stations and predicting how much power they will contribute to the mix. Often, trained meteorologists shift over to such work, which combines theoretical knowledge and field experience: another green-collar blend, in other words. Other desired backgrounds are computer science, aerodynamics, physics, and mathematics. Because of federal and state environmental-impact requirements, new sites must also be studied to determine how wind turbines might affect water supplies and plant and animal communities. This provides research and field opportunities for workers with bachelor's or graduate degrees in biology or environmental science.

Installing turbines by crane at the Horns Rev Offshore Wind Farm in the North Sea. *Photo: Elsam A/S (www.hornsrev.dk)*

Here are two listings for jobs within the broad field of assessment. As you'll see, both call for wide-ranging combinations of skills, experience, and education. The first is from Airtricity, a leading renewable-energy utility that is active throughout North American and the British Isles:

> This position reports to the WARS (Wind Analysis, Research & Standards) U.S. Manager and will be an active member of the WARS team: Early resource assessments and layouts; Occasional site visits to ascertain terrain, roughness and obstructions; Assistance with quality control and cleaning of large data sets from over 70 meteorological masts on a regular (3x weekly) basis; Assistance with delivery of high quality datasets of wind measurements from each site suitable for use in a low uncertainty wind resource assessment; Internal electricity production estimates using WAsP & WindFarmer, final layouts and management of subcontracted analyses; Internal site condition assessments; Use of GIS tools such as ArcView, Delorme XMap and GlobalMapper to produce high quality maps; Preparation of site reports; Implementing WARS safety standards for meteorological masts; Possibly overseeing power performance testing of wind turbines.
>
> Position Requirements: BS in Physical Science, Atmospheric Science, or Mechanical Engineering; Preferably 1–3 years post-qualification experience in the wind resource assessment field.

The second, at a higher level, is from Black & Veatch, a design and manufacturing firm. It seeks an experienced engineer to assist

A technician enters the tower base to climb up for the nacelle installation.
Photo: Warren Gretz, NREL.

with and manage all aspects of wind energy project assessment and development. Key responsibilities of this position include:

- ◆ Develop, plan, coordinate and manage all aspects of wind and other renewable energy projects.

- ◆ See projects from concept through completion. Areas of emphasis may include: Utility-scale wind farms, community wind systems, distributed wind, and offshore wind projects.

- ◆ The position will involve a wide variety of diverse assignments including: resource assessment, due diligence, siting, feasibility studies, new technology evaluation, project proposal evaluation, conceptual design, design review, performance estimates, and cost estimates.

- ◆ Interaction with clients including: utilities, energy project developers, financial institutions, energy end-users, investors, government agencies, and other organizations.

- ◆ Manage projects, including scope definition, project staffing, technical oversight, and responsibility for successful project execution on time and within budget.

- ◆ Lead and contribute to authorship of reports, studies, and proposals.

- ◆ Manage work of others in multi-disciplinary teams.

- ◆ Lead and assist with business development activities in area of expertise.

- ◆ Assist with development of new project opportunities and business lines.

Top: Preparing the tower base. *Photo: NEG-Micon* Bottom: Service work on a commercial turbine. *Photo courtesy of Vestas Wind Systems A/S*

- ◆ Exposure to wind resource analysis and production estimation. Although the primary focus of the job will be wind energy projects, the position would

2006 was a record-breaking year, with new installations of more than 2,400 MW. The new generating capacity installed in 2006 represented a capital investment of almost $4 billion, more than 10,000 new job-years nationwide (that is, 10,000 one-year jobs or 1,000 ten-year jobs), and more than $5 to $9 million in annual payments to landowners.
SOURCE: U.S. DEPT. OF ENERGY

include exposure to all renewable technologies: biomass, wind, geothermal, solar, hydro, etc.

The preferred candidate will possess:

- Engineering degree from an ABET-accredited university. Masters degree or MBA would be beneficial. Some course work on energy systems, specifically some knowledge of renewable energy technologies preferred.

- Minimum 8 years experience with wind energy projects is a necessity.

- Excellent verbal and written communication skills.

- Ability to communicate clearly and succinctly through formal reports, presentations, memoranda and email.

- Must be able to function equally well in collaborative, multi-discipline teams, and self-directed independent assignments.

- Must be self-motivated, with an ability to balance multiple projects while working under tight deadlines.

- Needs to have experience in wind energy projects specifically.

- Should have an interest in all renewable energy technologies.

- Experience with conceptual and detailed design is desired. Project field experience is also a plus.

Both positions, it's plain to see, require a solid educational background and significant experience in the field. Getting there is a challenge, but there are plenty of entry-level positions in every aspect of wind power that will set you on the right road.

The sector in wind energy that will produce the majority of new jobs in the next decade is manufacturing, installation, and operation; the demand for new plants is great, so much so that new clients often are put on a waiting list for the next available units. The manufacturing sector tends to hire trained engineers with degrees from four-year colleges, though two-year programs are increasingly contributing to the labor pool. A two-year background is sufficient, for instance, for workers who produce blades, towers, and gearboxes, the stuff of which wind towers are made; the control systems for them are electrical, therefore requiring training in electronics. Here, for example, is a job listing for a position as a "commissioning technician":

An international renewable energy equipment provider is seeking several Wind Turbine Commissioning Technicians for direct hire positions. Positions are available in PA, TX, & IL. Technician troubleshoots mechanical and electrical problems on variable-pitch, variable-speed turbines. Assists in all areas of site construction and commissioning as directed by the designated site supervision. Performs all mechanical and electrical component testing and ultimately repairs as needed. The position requires knowledge of direct current electricity/ alternating current electricity. The technician must be able to consistently climb wind turbine towers, to consistently travel for up to six weeks at a time between short trips home equaling two days and to work outdoors under extreme weather conditions. Extensive travel and overtime is required when needed to meet project schedules.

Lifting a nacelle onto the tower. *Photo courtesy of Vestas Wind Systems A/S*

**Top 12 States with
Wind Energy Potential**

1   North Dakota
2   Texas
3   Kansas
4   South Dakota
5   Montana
6   Nebraska
7   Wyoming
8   Oklahoma
9   Minnesota
10 Iowa
11 Colorado
12 New Mexico

The wind-power industry also offers ample opportunities for other field technicians, installation technicians, and maintenance workers. These jobs, many set at residential as well as commercial sites, have different requirements, some calling for one- or two-year certificates, others bachelor's degrees. At the most basic level of manufacturing and installation, a high-school diploma is sufficient, though these basic jobs tend to pay far less than other positions—no more than $15.00 an hour. Such positions have requirements of their own, of course, including, as the education ministry of the wind-rich Canadian province of Alberta puts it, "Skills in math, reading, and writing in order to read, interpret, and work from complex instructions, diagrams, and prints; computer literacy; capable of using and operating various equipment including torque wrenches and basic electrical test equipment (electrical or mechanical training is preferred); fiberglass repair skills; good physical condition (some physical activities include ladder climbing and heavy lifting); the ability to work at substantial heights (up to and exceeding 185 feet);

Installed U.S. Wind Energy Capacity

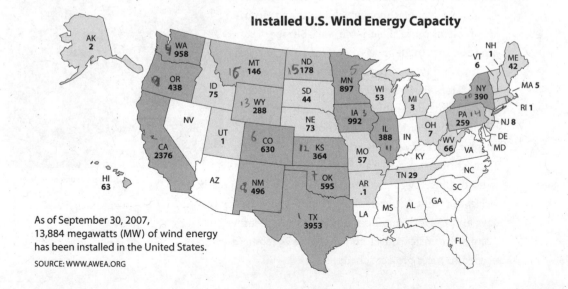

As of September 30, 2007, 13,884 megawatts (MW) of wind energy has been installed in the United States.

SOURCE: WWW.AWEA.ORG

problem solving, decision-making, and troubleshooting abilities; ability to work as part of a team as well as independently; must be self-motivated, energetic, and possess a positive attitude; a commitment to safe work habits; safety certification."

Knowledge Is Power

Many positions in wind energy research require an engineering background and a four-year degree in some aspect of engineering. Others, as we have seen, involve environmental sciences, meteorology, physics, mathematics, and business. The most capable players in the field, those who are likeliest to succeed in the coming years as the business grows and changes, will combine technical abilities, theoretical knowledge, people skills, business smarts, and a hunger to know all there is to know about renewable energy.

Several degree-granting programs offer training in wind energy—Texas Tech University Wind Science and Engineering Research Center, the University of Massachusetts Center for Energy Efficiency and Renewable Energy, the University of Utah Wind Energy Research Program, the MIT Laboratory for Energy and the Environment, and the Illinois Institute of Technology Energy and Sustainability Institute among them—and many more offer laboratory programs that allow for guided independent study in the area.

Several community colleges offer programs leading to an associate degree or certification. These programs are not yet as widely developed as those for solar energy, but their numbers are growing. One particularly interesting program, and a model for others, is offered by the Iowa Lakes Community

The cost of producing electricity from wind power has dropped from 80 cents per kilowatt-hour in 1980 (in current dollars) to 4 to 6 cents in 2007.
SOURCE: U.S. DEPT. OF ENERGY

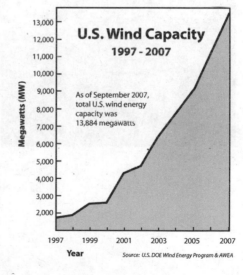

U.S. Wind Capacity
1997 - 2007

As of September 2007, total U.S. wind energy capacity was 13,884 megawatts

Megawatts (MW)

Year

Source: U.S. DOE Wind Energy Program & AWEA

College. The ILCC Program in Wind Energy and Turbine Technology graduates about thirty students a term, and almost all of them enter the industry immediately. Taking the first three terms leads to a certificate; completing the whole course of 80 units leads to an associate's degree in applied science.

ILCC Program in Wind Energy and Turbine Technology

First Term
Field Training and Project Operations
Direct Current Electrical Theory
Alternating Current Electrical Theory part I
Introduction to Wind Energy
Introduction to Computers/Info Systems
Intermediate Algebra

Second Term
Human Relations
Alternating Current Electrical Theory part II
Wind Turbine Mechanical Systems
Basic Hydraulics
Business Communications
Electric Motors and Motor Controls Fundamentals

Third Term
Wind Turbine Site Construction / Locations
Wind Turbine Internship

Fourth Term
Power Generation and Distribution
Basic Networking and Computer Technology
Meteorology (or Chemistry or Physics)
Electrical Practical Applications
Airfoils & Composite Repair

Photo © Bonus

Fifth Term
Data Communication and Acquisition
Programmable Logic Control Systems
Wind Turbine Siting
Principles of Management

Apprenticeship and internship is another good avenue into a career in wind energy. Visit the Wind Energy Career Center at the AWEA web site (see the resources below) for a list of available internships and job openings. At all levels of education and experience, it's a good idea for professionals to give themselves continuing education through workshops, seminars, and conferences, also listed on the AWEA site. ❖

MEDIAN SALARIES/WAGES ❖ Wind Energy		
Electrical Engineer	$55,040	– $96,410
Mechanical Engineer	$52,230	– $85,320
Material Engineer	$54,980	– $90,219
Aerodynamics Engineer	$62,680	– $105,790
Atmospheric Scientist	$58,240	– $94,910
Power Distribution Engineer	$51,490	– $82,300
Wind Turbine Installer	$16.00 – $21.90 / hour	
Meteorologist	$19.50 – $29.70 / hour	

SOURCE: WWW.PAYSCALE.COM (DECEMBER 2007)

Wind Energy Resources

ORGANIZATIONS & WEB SITES

American Wind Energy Association
AWEA is a trade and advocacy organization based in Washington, D.C. It represents the U.S. wind energy industry, sponsoring an annual national conference devoted solely to wind energy, which features technical papers from experts in all areas of the field. Visit their Job Board for a wealth of information.

1101 14th Street NW, 12th Floor Washington, DC 20005
(202) 383-2500
www.awea.org

European Wind Energy Association
EWEA is the voice of the wind industry in Europe and worldwide. Its members, from 40 countries, include over 300 companies, associations, and research institutions. These members include manufacturers covering almost all of the world wind-power market, component suppliers, research institutes, contractors, national wind and renewables associations, developers, electricity providers, finance and

insurance companies and consultants. The EWEA web site is full of useful information on every aspect of wind power.

www.ewea.org

Great Plains Windustry Project

GPWP offers an information-rich site with advocacy for community wind projects, which are locally owned by farmers, investors, businesses, schools, utilities, or other public or private entities. The key feature, they note, is that local community members have a significant, direct financial stake in the project beyond land-lease payments and tax revenue.

> 2105 1st Avenue South
> Minneapolis MN 55404
> (800) 946-3640 or
> (612) 870-3461
> *www.windustry.org*

U.S. Department of Energy Wind and Hydropower Technologies Program

This program is leading the nation's efforts to improve wind energy technology so that it can generate competitive electricity in areas with lower wind resources, and to develop new, cost-effective, advanced hydropower technologies that will have enhanced environmental performance and greater energy efficiencies.

> *www1.eere.energy.gov/wind andhydro/*

PUBLICATIONS

Paul Gipe, ***Wind Power Renewable Energy for Home, Farm and Business*** (Chelsea Green, 2004) Includes chapters on how to evaluate modern wind turbine technology, installing wind turbines safely, designing a stand-alone power system for living off-the-grid, and how to use electricity-producing wind turbines to pump water in rural areas.

J. F. Manwell, J. G. McGowan, and A. L. Rogers, ***Wind Energy Explained*** (Wiley, 2002) A comprehensive survey of wind energy, presupposing a solid basic understanding of mathematics and physics.

Sathyajith Mathew, ***Wind Energy: Fundamentals, Resource Analysis and Economics*** (Springer, 2006) A thorough analysis of wind-energy economics, with an emphasis on material of great use to "green-collar" workers whose work blends technical and nontechnical subjects.

Wind Power Monthly
A source of independent international news about developments in wind and other renewable energies. Subscribers receive access to industry statistics not easily found elsewhere.

> *www.windpower-monthly.com*

Geothermal

3 Geothermal Energy

Of all the renewable sources of energy that are available to us, the one that may be the most easily accessible lies beneath our feet. Heat from the earth, called geothermal energy, uses steam and warm water that rise naturally from underground reservoirs. Across much of the planet, the subsurface temperature rises substantially—by as much as 150°F (66°C) in some places—for every mile of descent, so that a pipe run to less than a mile can bring up water warm enough to bathe in and wash the dishes. One method now being explored is to send cold water to a depth of about two miles in an especially hot zone and let the earth heat it naturally; when the water returns to the surface, it does so at a temperature of 460 to 600°F (237–315°C), enough to create plenty of electricity from steam. This indirect use of geothermal energy is very promising: there's plenty of hot underground water available, enough to supply our energy needs for 100,000 years. Getting to it is a challenge, but then, so is coming up with energy solutions of every other kind, too.

Old Faithful is a testament to the power of geothermal energy. *Photo: Gregory McNamee*

The water need not even be especially warm to be usable. Just a hundred feet below the streets of Chicago, the water flows at a comfortable 55–60°F (13–16°C). Tapping into it by way of a geothermal heat pump (GHP) brings up water that needs to be heated only a few degrees for use in heating in winter, and that's a perfect temperature to send through flooring and ceiling pipes to cool buildings in summer. These medium-temperature water reservoirs can be used in many ways without much modification, providing direct use of geothermal energy. They are abundant in the Midwest, where GHP technology and other geoexchange systems are most widely employed at present. Overall, geoexchange systems are more energy-efficient

How Geothermal Energy Works: Geoexchange

Geoexchange (sometimes called geothermal, or ground-source heating and cooling) taps the renewable, safe, and virtually endless energy supply that lies just below the earth's surface. In winter, warmth is drawn from the earth through a series of pipes, called a loop, installed beneath the ground. A water solution circulating through this piping loop carries the earth's natural warmth to a heat pump inside the home. The heat pump concentrates the earth's thermal energy and transfers it to your home.

In the summer, the process is reversed; heat is extracted from air inside the house and transferred underneath the surface by way of the ground loop piping. The geoexchange system also uses some of the heat extracted from the interior in the summer to provide free hot water—saving as much as 30 percent on your annual hot water bill.

Because geoexchange technology uses such a readily available source of energy—and uses it so efficiently—it can save a substantial amount of money on monthly utility bills. In fact, a typical 1,500-square-foot home in a moderate climate can be heated and cooled for a year-round average of just $1 a day.

SOURCE: GEOTHERMAL HEAT PUMP CONSORTIUM

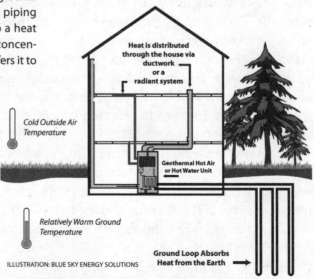

Heat is distributed through the house via ductwork or a radiant system

Cold Outside Air Temperature

Geothermal Hot Air or Hot Water Unit

Relatively Warm Ground Temperature

ILLUSTRATION: BLUE SKY ENERGY SOLUTIONS

Ground Loop Absorbs Heat from the Earth →

than other systems, which translates into lower energy bills and a reduction in the consumption of fossil fuels. They are also clean. Lake County, California, the largest geothermal field under production in the world, is one of the few counties in the state to meet all federal and state air-quality standards. The county's air quality has improved in recent years, in fact, because hydrogen sulfide that would ordinarily be released naturally into the atmosphere by hot springs and vents is now filtered out at the geothermal plants—reducing hydrogen sulfide emissions by a whopping 99.9 percent!

Geothermal, in short, has everything going for it. Even though the technology can be expensive, the returns come immediately, making it a comparatively easy sell for those whose jobs it is to—well, sell it.

Even so, countries such as Japan, Iceland, New Zealand, Italy, Indonesia, and Canada have been at the forefront of geothermal-energy development to date. Though the United States has vast geothermal resources, only one percent of the power used in this country is so produced. As of 2007 the geothermal industry in the U.S. has a capacity of nearly 3,000 megawatts (MW), with power plants operating in five states employing about 5,000 people. According to the Geothermal Energy Association, over 34,000 jobs are expected to result from a doubling of that capacity, with manufacturing and construction jobs typically corresponding to 6.4 jobs per MW of capacity and 0.74 permanent full-time jobs per MW directly related to power plant operation and maintenance. "If the additional jobs brought on by research, direct use applications, and other geothermal activities are considered, the number of direct jobs would be even greater," adds the GEA, noting that these figures consider employment created by electricity production alone.

These figures are conservative overall. The GEA estimates that

A worker monitors equipment at The Geysers geothermal power plant, one of the largest in the world. *Photo: David Parsons, NREL*

with effective federal and state support, as much as 20 percent of the nation's power needs can be met by geothermal energy sources by 2030. Analysts expect the contribution of geothermal to grow markedly in the coming years: the energy is abundant, easily obtained, and, best of all, inexpensive in the long run compared to most other sources of power. That growth means jobs and many opportunities.

Careers in Geothermal Energy

As with other branches of the renewable-energy industry, geothermal energy requires the services of skilled workers, professionals, salespeople, and auxiliary staff such as secretaries, accountants, attorneys, surveyors, and warehouse workers.

Geothermal pumps can be used nearly worldwide. The earth's temperature a few feet below the surface is a relatively constant temperature of about 45–58 degrees F, while air temperatures can fluctuate greatly.

Identifying underground reservoirs is the work of trained geologists, geochemists, hydrologists, and geophysicists, almost all of whom enter the industry after undergraduate and, often, graduate training in the geosciences. Hydraulic engineers, often with training in civil engineering as well as geosciences, are responsible for overseeing the drilling of those reservoirs. Environmental engineers and biologists are charged with preparing environmental-impact studies of areas under consideration for development. Mechanical and electronic engineers, chemists, and materials scientists are also involved in researching and developing new and improved geothermal-energy technologies.

In constant demand are geophysicists, who specialize in exploration techniques that use seismic, electromagnetic, and other sensing methods in the search for oil, gas, minerals, water, and the like. Says one job description: "A geophysicist interprets seismic data, and recommends

Pipelines to a geothermal plant.

drilling prospects and processing techniques. A variety of equipment and modeling methods are used to prepare maps (structure, contour, isopach and others) that provide essential information for reservoir development and forecasting. The data is also used to support plant operations and comply with governmental reporting." An entry-level position requires at least a bachelor's degree in physics or geophysics, with graduate training preferred.

The direct-use technologies employed by geothermal energy require workers trained in heating and air-conditioning systems, as well as in the building trades. The interface with geothermal energy and electrical systems requires electrical technicians, electricians, electrical machinists, welders, mechanics, and other skilled workers involved in constructing, operating, and maintaining power plants.

Mechanical engineers, geologists, drilling crews, and heating, ventilation, and air conditioning contractors are needed to manufacture and install GHPs. (Less maintenance is required for geoexchange systems than other technologies, since most of the equipment involved is not exposed to the elements.) At a specialized level, a geothermal heat pump (HVAC) engineer troubleshoots, repairs, and maintains the whole system: pipes, motors, electrical components, generators, and power transfer switches. Here is a job description posted on one site in the so-called Tech Valley of upstate New York, where geothermal energy is now being developed:

Amédee Geothermal Venture I power plant in Wendel, California.
Photo: J.L. Renner, INEEL

A Geothermal Heat Pump HVAC Engineer must continuously evaluate HVAC operation, performance and documentation requirements, identify specific improvement needs, and provide recommendations or implement system upgrades or improvements. The Engineer may be

A drilling company drills two 250-foot holes for a geothermal loop. Another local company will install the geoexchange equipment for a super-efficient 1,450 square-foot home in northern California. *Photo courtesy of Doug Pratt*

required to coordinate contractor personnel who perform repairs, modifications, and installation of HVAC and refrigeration equipment. He or she must have the ability to work on any repair or project as needed, independently or as a team member. He or she must work with a variety of hand tools, measuring devices, power tools, milling machines, lathes, band saws, welding equipment, bench grinders, and materials. The ability to perform minor electrical repairs, along with ability to read electrical schematics and to detect and report improper operations, faulty equipment and defective machines and unusual conditions to proper supervision is also required. He or she must perform repairs to AC systems and air handlers and create and maintain accurate service logs for a variety of building equipment including water source heat pumps, cooling tower, and other equipment.

GHP engineers, the site goes on to remark, are "in high demand and can expect large career growth possibilities. With a degree and experience, management and supervisory positions can be obtained." Entry-level positions start at around $30,000; a worker with five years of experience can expect to make $40,000–$50,000.

Knowledge Is Power

The preceding position for a GHP engineer requires a high-school diploma or the equivalent and appropriate HVAC certifications. "A two-year degree and bachelor's degrees for advancement/supervisory positions may be preferred but not always required," the advertisement continues.

Other positions within the geothermal-energy field require a degree from a community college or four-year college. One especially promising program, inaugurated in the fall of 2007, is located at the University of Nevada at Reno, where, in concert with the Truckee Meadows Community College, students have the vast Steamboat Geothermal Complex nearby as a field laboratory.

Workers within the field can benefit from workshops and conferences (see resources at the end of this chapter), as well as advanced training. One new development is the Association of Energy Engineers (AEE) Certified GeoExchange Designer (CGD) program, which, AEE says, is "designed to recognize professionals who have

Blue Sky Energy Solutions

David Petroy and business partner Eldred Himsworth design commercial and residential geothermal systems in Colorado (*www.bluesky-energy.com*). David shares his thoughts:

What motivated you to launch Blue Sky Energy Solutions? *I've been involved in energy industry throughout my professional career and have always had a personal passion for the outdoors and nature. Launching a company that helps society adopt new and better ways to provide the power needed to function satisfies my scientific and engineering interests in energy and helps lessen the impact of fossil fuels on the environment.*

For people wanting to enter this field, what education and skills do you recommend? *For system designers I would recommend a mechanical engineering degree or a building-energy engineering degree. For applications engineers, good communication skills are essential to understanding the client's needs and for presenting appropriate options. Both designers and engineers should study energy economics and/or micro economics, and a course or two in hydrogeology and/or near-surface geology would be very helpful. And everyone should have a basic course in business.*

For installers, certification in excavating equipment operation and a commercial truck license is required. We will need many more well drillers if the industry is going to expand. This is a specialized field which requires drilling-training certification and apprentice experience for a few years.

Do you see growth potential in this industry? *The geothermal (ground source heat pump) heating industry has been growing slowing and steadily over the past 20 years through all types of economic conditions. It is a well-proven, solid technology. The work is rewarding and every job has a slightly new twist—which keeps things interesting.*

demonstrated high levels of experience, competence, proficiency, and ethical fitness in applying the principles and practices of geothermal heat pump design and related disciplines, as well as to raise the professional standards within the field, and to encourage those involved in the design process through a continuing education program of professional development." Eligibility for the program is contingent on meeting one of the following sets of requirements:

A four-year engineering degree and/or P.E. [Professional Engineer], and/or R.A. [Registered Architect] with at least three years of combined experience in commercial geothermal heat pump design, and/or heating, ventilating and air conditioning field.

A four-year non-technical degree, with at least five years of combined experience in commercial geothermal heat pump design, and/or heating, ventilating and air conditioning field.

A two-year technical degree, with at least eight years of combined experience in the geothermal heat pump design, heating, ventilating and air conditioning field.

MEDIAN SALARIES/WAGES ❖ Geothermal		
Geologist	$42,830	– $82,650
Senior Geologist	$60,500	– $140,500
Hydrologist	$44,110	– $71,220
Geoscientist	$55,500	– $106,950
Geophysicist	$45,740	– $134,880
Hydraulic Engineer	$54,040	– $76,480
Environmental Engineer	$55,590	– $73,660
HVAC Mechanic	$16.00 – $25.00 / hour	
HVAC Field Service Engineer	$23.90 – $30.60 / hour	
SOURCE: WWW.PAYSCALE.COM (DECEMBER 2007)		

Ten years or more of verified combined experience in commercial geothermal heat pump design, and/or heating, ventilating and air conditioning field.

All candidates for certification must take an online exam. Those who pass the test can expect many opportunities for professional advancement—and lots of work in the years to come. But at every level—white collar, blue collar, and green collar—geothermal energy holds plenty of promise for everyone. ❖

Geothermal Resources

ORGANIZATIONS & WEB SITES

Association of Energy Engineers is a nonprofit professional society of 8,500 members in 77 countries. Its mission is to promote the scientific and educational interests of those engaged in the energy industry and to foster action for sustainable development. The society publishes useful information on its web site, and it offers a range of scholarships in energy and management.

> 4025 Pleasantdale Rd., Suite 420
> Atlanta, GA 30340
> (770) 447-5083
> *www.aeecenter.org*

Geothermal Education Office GEO produces and distributes educational materials about geothermal energy to schools, energy/environmental educators, libraries, industry, and the public. GEO collaborates frequently with education and energy organizations with common goals, and, through its web site, responds to requests and questions from around the world.

> 664 Hilary Drive
> Tiburon, CA 94920
> (415) 435-4574
> *geothermal.marin.org*

Geothermal Energy Association is a trade association composed of U.S. companies that support the expanded use of geothermal energy and are developing geothermal resources worldwide for electrical power generation and direct-heat uses. Its web site offers a wealth of information on all aspects of geothermal energy, as well as information on its annual trade show, which draws exhibitors and participants from all over the world.

> 209 Pennsylvania Avenue SE
> Washington, DC 20003
> (202) 454-5261
> *www.geo-energy.org*

Geothermal Heat Pump Consortium Another trade association, the GHPC is a first-rate source of information about geoexchange technologies, professional licensing, and other topics.

> 1050 Connecticut Ave. NW
> Suite 1000
> Washington, DC 20036
> (202) 558-6335
> *www.geoexchange.org*

Geothermal Resources Council GRC members represent a broad spectrum of geothermal professionals and companies from around the world. They serve as the focal point of continuing professional development for its members through many outreach, information and technology transfer, and educational services. Membership is open to any person, company, organization or agency. Their web site also provides easy access to the International Geothermal Association.

> P.O. Box 1350
> Davis, CA 95617
> (530) 758-2360
> *www.geothermal.org*

Geothermal Technologies Program, U.S. Department of Energy
This DOE site offers news on educational programs, scholarships, and workshops as well as documents of many kinds on geothermal energy. *www1.eere.energy.gov/geothermal*

PUBLICATIONS

Mary H. Dickson and Mario Fanelli, eds., *Geothermal Energy: Utilization and Technology* (Earthscan, 2005)
Beginning with an overview of geothermal energy and the state of the art, leading experts in the field cover direct use of geothermal energy, electricity generation, environmental impact, and economic, financial and legal considerations. Case studies are included throughout.

Harsh K. Gupta and Sukanta Ray, *Geothermal Energy: An Alternative Resource for the 21st Century* (Elsevier, 2006)
This book studies various facets of geothermal energy development and summarizes the current knowledge of geothermal resources and their exploitation. Accounts of geothermal resource models, various exploration techniques, and drilling and production technology are discussed.

National Renewable Energy Laboratory, *21st Century Complete Guide to Geothermal Energy, Geothermal Heat Pumps, Electricity, Potential, Drilling, Photo Gallery, Geopowering the West* (NREL, 2004)
This electronic book on CD-ROM contains federal documents on energy from various agencies, including the Department of Energy (DOE) Energy Efficiency and Renewable Energy Network (EREN) Geothermal Energy Program; National Renewable Energy Laboratory (NREL); GeoPowering the West; U.S. Geothermal Potential; and DOE Climate Challenge.

Karl Ochsner, *Geothermal Heat Pumps: A Guide for Planning and Installing* (Earthscan, 2007)
Ochsner, a European specialist in geothermal systems, introduces basic theory and reviews the wide variety of available technology for homes and businesses. The book offers information on planning and system control, using data, graphics, and tables from the worldwide market.

4

Hydroelectric Power & Marine Energy

4   Hydroelectric Power & Marine Energy

You hear it long before you see it, a dull roar shrouded in steam, hidden from view. Then you see it: Horseshoe Falls, where the Niagara River drops 180 feet (55 meters) into Niagara Gorge, on the border between the United States and Canada. Between May and November, when the river is usually free of ice, 675,000 gallons of water tumble over the falls each and every second.

Niagara Falls
Photo: Gregory McNamee

On the other side of the country, the Pacific Northwest has long relied on water-generated electricity. Today there are some 160 hydroelectric facilities in the region, providing Washington State with nearly 80 percent of its power and Idaho nearly all of its supply. The biggest of these facilities is the Grand Coulee Dam. California and Oregon also have numerous facilities.

Hydroelectricity is the most prevalent form of renewable energy throughout the nation, accounting for about 90 percent of the total of renewable resources. Most of the nation's dams—and there are

Hydropower does not produce waste, greenhouse gases, or other air pollution.

about 80,000 of them—do not generate electricity, but many are now being studied for the possibility of retrofitting them to produce power. Added to the mix of power, hydroelectricity is clean and abundant. It is also inexpensive relative to other forms of power, and as long as dams and plants are engineered in such a way that the health of rivers and their wildlife populations is assured, there are few reasons not to put water to greater use in providing power, at least in places where water is abundant.

In that spirit, current engineering research is focusing on building turbines that do not require damming and that allow fish to pass through them without harm. There is also exciting research being

How Hydropower Works

Mechanical energy is derived by directing, harnessing, or channeling moving water. The amount of available energy in moving water is determined by its flow or fall. Swiftly flowing water in a big river, like the Columbia River along the border between Oregon and Washington, carries a great deal of energy in its flow. So, too, does water descending rapidly from a very high point, like Niagara Falls in New York. In either instance, the water flows through a pipe, or penstock, then pushes against and turns blades in a turbine to spin a generator to produce electricity.

In a run-of-the-river system, the force of the current applies the needed pressure, while in a storage system, water is accumulated in reservoirs created by dams, then released when the demand for electricity is high. Meanwhile, the reservoirs or lakes are used for boating and fishing, and often the rivers beyond the dams provide opportunities for whitewater rafting and kayaking. Hoover Dam, a hydroelectric facility completed in 1936 on the Colorado River between Arizona and Nevada, created Lake Mead, a 110-mile-long national recreational area that offers water sports and fishing in a desert setting.

SOURCE: U.S. DEPARTMENT OF ENERGY

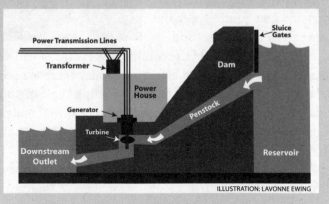

ILLUSTRATION: LAVONNE EWING

undertaken on a still larger, inexhaustible source of power—namely, oceanic tides, which can turn massive turbines and provide abundant energy to vast portions of the world. Only a couple of dozen places on Earth have natural inlets and a large enough tidal range—about 10 feet (3 meters)—to produce energy economically, but that has more to do with our technology than with the potentials of nature. That technology is still very much in its early stages. There are currently two commercial tidal-energy dams in operation, one in France, the other in Canada; energy economists have identified several other potential sites in England and Russia.

Tidal and wave energy holds extraordinary promise to put the bounty of the earth to good use without causing harm. As with all forms of renewable energy, water power, from the oceans and from the rivers alike, needs the minds and efforts of students, researchers, thinkers, and workers from all places and many walks of life.

Careers in Hydropower

As with other forms of renewable energy, hydroelectric power requires a blend of skilled workers, technicians, researchers, operators, engineers, electricians, salespersons, and auxiliary staff. Projected hydroelectric plants must also be assessed for their potential effects on fish and wildlife populations and on river flows, affording opportunities for wildlife specialists, environmental scientists, biologists, hydrologists, and other scientists. Arguments for the construction of new projects—pro and con—require advocates, writers, and legal specialists. Economists and energy analysts study the economic need for new facilities and

Over one-half of the total U.S. hydroelectric capacity for electricity generation is concentrated in three States: Washington, California and Oregon.
SOURCE: ENERGY INFORMATION ADMINISTRATION

Ice Harbor Dam near Burbank, Washington.
Photo: US Army Corp of Engineers

Hydropower produces enough electricity to serve the needs of 28 million residential customers in the U.S.—equal to all the homes in Wisconsin, Michigan, Minnesota, Indiana, Iowa, Ohio, Missouri, Nebraska, Kansas, North and South Dakota, Kentucky, and Tennessee.

their effects on business and society. In short, hydroelectric power welcomes individuals with many skills and interests.

At the heart of hydroelectric energy development is hydrology, the scientific study of water and its properties. Hydrologists are called on to analyze where water supplies can be found and how long those supplies are likely to last, and no discussion of resource policy is complete without their participation. Early in their careers, hydrologists are likely to spend most of their time in the field, mapping water supplies and pathways, testing water quality, and planning projects on the ground. As they advance, many hydrologists engage in laboratory or office work, often acting as liaisons to government agencies and businesses and analyzing computer-generated projections of water availability and behavior.

According to the U.S. Bureau of Labor Statistics, in 2004 there were about 8,000 positions for hydrologists at all levels in the United States, with a median annual income of $61,510. It was projected that the demand for hydrologists would grow slightly throughout the rest of the decade, mostly with an eye toward monitoring compliance with environmental regulations, as well as working in flood control. Increased emphasis on hydroelectric power would increase opportunities for hydrologists.

Hydraulic engineers and hydraulic technicians work in conjunction with hydrologists to develop and operate hydroelectric plants and related facilities. Civil engineers and structural engineers are involved in building dams and other structures, and electrical engineers and mechanical engineers develop and maintain power systems that are in turn staffed by power-plant operators and other technicians. Most such systems are operated and monitored by computer, requiring skilled information-technology workers.

Additionally, many dams and hydroelectric facilities are located within federal parks, national forests, and other parts of the public domain, served by park rangers, forest-service officers, and other officials. In many such venues, there are also opportunities for recreation workers, wilderness guides, and other outdoor-activities specialists.

Knowledge Is Power

Most technical positions in hydroelectric power require an associate's degree at minimum. Here, for example, are the posted requirements for a senior hydroelectric power technician whose principal duties are to repair and maintain the hydraulic, electrical, and mechanical systems of a power plant in the Northeast:

◆ Must have a minimum of 7 years experience working with complex mechanical, hydraulic, and electrical systems—ideally including turbines.

◆ Must be able to read and understand electrical, hydraulic, and mechanical drawings and schematics.

◆ Must have the ability to work well on a team basis.

Ocean Power Technologies' PowerBuoy® is designed to convert ocean wave energy into useable electrical power for utility-scale grid connected applications. *Photo: Ocean Power Technologies (www. oceanpowertechnologies.com)*

The United States is the second-largest producer of hydropower in the world, behind Canada. Norway produces more than 99 percent of its electricity with hydropower. New Zealand uses hydropower for 75 percent of its electricity.

◆ Strong verbal and written communication skills are essential.

◆ Solid problem-solving and organizational ability is a must.

◆ Must have familiarity with OSHA safety standards.

◆ Knowledge of basic Windows software including Outlook, Excel, Word, and Access is required.

◆ Must have a minimum of a two-year degree in Electrical Engineering Technology or a related technical discipline.

The successful candidate for such a job might find himself or herself working alongside or even for a project engineer, whose skill set and education will be substantially different. Here is a posting for that position in the same geographic area:

Work will involve leading and performing the preparation of conceptual, feasibility and preliminary and final design work products. The work requires coordination with clients, self direction and ability to work as an individual or in a group. Activities will include a wide range of water resources development study tasks, such as hydrological studies to estimate river flows and floods, formulation of dam and hydropower project development alternatives, preparation of preliminary layouts of the major project components, estimating quantities of construction materials, assisting in preparation of project cost estimates, assisting in operational studies to evaluate river flow and hydropower production, and economic studies to evaluate alternative projects and to optimize overall project configuration. Activities will be focused on civil engineering project development aspects, in coordination with mechanical and electrical specialists. The work environment is results oriented and focused on preparing reports or other deliverables within established budgets and schedules. Hands-on knowledge and familiarity with modern computation hydrologic and hydraulic computer methods and CAD systems are required.

Electricity from the Ebb and Flow of Tides

Verdant Power's Roosevelt Island Tidal Energy (RITE) Project is being operated in New York City's East River. The RITE Project incorporates a Kinetic Hydropower System comprised of Verdant Power's 5 meter, 35 kW tidal Free Flow™ turbines, which generate electricity from the ebb and flow tides of the East River. The Project is progressing from an initial demonstration array of six turbines to a complete arrangement of 100-300 tur-

bines. At full capacity the project could generate up to 10 MW, enough to power nearly 8,000 New York homes. *www.VerdantPower.com*

Top: RITE Project Phase 2 – A Verdant Power Free Flow turbine is being installed into the East River, New York City. Left: Installation of fish monitoring equipment. *Photos courtesy of Verdant Power, Inc. (www.VerdantPower.com)*

Hydropower is the most efficient way to generate electricity. Modern hydro turbines can convert as much as 90 percent of the available energy into electricity. The best fossil fuel plants are only about 50 percent efficient.

That job, the advertisement continues, requires a master's of science in civil engineering with an emphasis on hydropower planning, hydrology, hydraulics, and sedimentation, as well as professional engineering registration and, as with the first position, a minimum of seven years' experience.

A beginning hydrologist, in most instances, must hold a bachelor's degree from a scientific program. Typically, a hydrologist will have training in geophysics, chemistry, engineering science, soil science, mathematics, computer science, aquatic biology, atmospheric science, geology, and oceanography—not all of these fields, of course, but a number of them in combination. Economics, public finance, public policy, environmental law, and business law provide helpful background for hydrologists at more senior levels, and it helps to be comfortable writing reports and giving talks or testimony about water issues.

On that note, here is a job listing for a position as a hydropower engineer working for the Federal Energy Regulatory Commission:

As an entry-level staff engineer you would work with an interdisciplinary team of environmental scientists and engineers, to review, analyze, and resolve engineering and environmental issues associated with proposals to construct and operate non-federal hydroelectric projects; including major dams, reservoirs and powerplants.

You would identify and evaluate alternatives to the project proposed and prepare written analyses describing the impacts of the projects and alternatives with emphasis on engineering-related issues including the need for project power, cost/benefit analysis, hydraulic and hydrologic modeling and the interrelationship of project facilities and flows with other environmental resources such as aquatics and recreation. You would propose recommendations to protect and enhance the environment.

You would visit the project locations before authorization to familiarize yourself with the project site, and if the project is existing, to become familiar with its current and proposed operation.

Your position would involve participation in public scoping meetings, and technical sessions with project proponents, state and federal resource agencies, and other affected parties. Some travel would be required. Training opportunities are provided and encouraged.

Qualifications: A bachelor's degree in engineering; Strong verbal, analytical and writing skills; U.S. Citizenship.

Many hydroelectric facilities within the United States are operated by the U.S. Army Corps of Engineers, which offers civilian employees a four-year training program. One instructor in that program suggests that students interested in such a career path study math and physics—but that they also get a good general education to help them with critical-thinking and problem-solving skills. ❖

Hydropower prevents the burning of 22 billion gallons of oil or 120 million tons of coal each year.

MEDIAN SALARIES/WAGES ❖ **Hydropower**	
Hydrologist	$44,110 – $71,220
Hydraulic Engineer	$54,040 – $76,480
Environmental Engineer	$55,590 – $73,660
Civil Engineer	$44,010 – $78,400
Electrical Engineer	$55,040 – $96,410
Mechanical Engineer	$52,230 – $85,320
Plumbing Engineer	$53,320 – $68,250
Civil Engineering Technician	$15.40 – $22.70 / hour
Pipefitter, Steamfitter	$16.90 – $36.70 / hour

SOURCE: WWW.PAYSCALE.COM (DECEMBER 2007)

Hydropower Resources

ORGANIZATIONS & WEB SITES

National Hydropower Association
Since 1983, NHA has been dedicated exclusively to advancing the interests of hydropower energy in North America. Its web site has a wealth of information, and well-tended links to other resources. The associated **Hydro Research Foundation** web site (*www.hydrofoundation.org*) has good information for students, too.
www.hydro.org

Ocean Energy, California Energy Commission
An excellent source for links to companies and organizations involved in ocean energy development.
www.energy.ca.gov/development/ oceanenergy/

U.S. Bureau of Reclamation
This web site is devoted to Hoover Dam, one of the nation's best-known hydroelectric facilities. It offers information on dam operation, as well as resources for students and teachers.
www.usbr.gov/lc/hooverdam

U.S. Department of Energy Wind and Hydropower Technologies Program
Working with industry, the Wind and Hydropower Technologies Program pursues R&D to develop more environmentally friendly technologies to maintain the nation's existing hydropower capacity.
www1.eere.energy.gov/wind andhydro/

PUBLICATIONS

William P. Creager and Joel D. Justin, ***Hydroelectric Handbook*** (Wiley, 1952)
Though more than half a century old, this is a widely used reference in the field, covering the theory and practice of hydroelectric power generation.

Scott Davis, ***Microhydro: Clean Power from Water*** (New Society Publishers, 2004)
Microhydro is the first complete book on the topic in a decade. It covers principles, design and site considerations, equipment options, as well as legal, environmental, and economic factors.

21st Century Complete Guide to Hydropower: Hydroelectric Power, Dams, Turbines, Safety, Environmental Impact, Microhydropower, Impoundment, Pumped Storage, Diversion, Run-of-River (Progressive Management, 2006)
This CD-ROM gathers important federal documents and resources on hydropower and related technologies such as dams and turbines, as well as material from the Federal Energy Regulatory Commission (FERC) regarding licensing, compliance, safety, and inspections.

Bioenergy

5 Bioenergy

Organic matter, or biomass, stores energy, and people have known this fact for a very long time. A tree, for instance, can be thought of as sunlight encased in wood, and the energy it contains—called bioenergy—has been a source of warmth and light for as long as there have been humans. Wood continues to heat our homes, and about one percent of the nation's energy supply comes from the burning of wood waste and other biomass. But things are on the move, and bioenergy shows much promise of becoming one of the leading kinds of renewable energy on the market.

In recent years, owing to state and federal clean-air regulations and other incentives, the petroleum industry has increasingly added bioenergy to nonrenewable fossil-fuel energy, in the form of ethanol. This renewable source, derived mostly from corn in the United States, is far from perfect: at present, under current means of production, it takes lots of fossil fuel to produce ethanol, and, according the Renewable and Appropriate Energy Laboratory at the University of California, ethanol provides only an 18 percent reduction in greenhouse-gas emissions over gasoline.

Current ethanol production is primarily from the starch in field corn kernels, but NREL researchers in the DOE Biofuels Program are developing technology to produce ethanol from the fibrous material in the corn stalks and husks, or other agricultural or forestry residues. *Photos: Warren Gretz, NREL*

Other sources are more efficient. Brazil, for instance, has been producing fuel from less costly, more sustainable sugarcane since the 1970s and has attained its goal of achieving energy independence in the bargain; Brazilian cars run on what is locally called *álcool*, which has the added virtue of having an octane rating of 113, practically putting every driver behind the wheel of a race car, as well as reducing greenhouse gas emissions by up to 70 percent. Trees and grasses, which are called cellulosic biomass, can also be used to make bioethanol, and some fast-growing varieties such as switchgrass, bamboo, hemp, and kenaf are particularly useful—and reduce greenhouse gas emissions to almost zero. Other sources include agricultural residues such as the stalks, leaves, and husks of corn plants; forestry wastes such as chips and sawdust from lumber mills, dead trees, and tree branches; municipal solid waste, including household garbage and paper products; and food processing, papermaking, and other industrial wastes.

A pile of biomass wood chips wait to be gasified at the McNeil Generating Station in Burlington, Vermont.
Photo: Warren Gretz, NREL

Another biofuel—biodiesel—is a mixture of fatty-acid alkyl esters made from vegetable oils, animal fats, and recycled greases. Biodiesel can be used directly in modified diesel engines (and in some newer, unmodified diesel engines), but it is usually used as a petroleum diesel additive. Biodiesel burns fairly cleanly, even if people sometimes joke that vehicles that use it smell like french fries. It burns so cleanly, in fact, that it promises to reduce airborne particulates, carbon monoxide, and hydrocarbons significantly in the coming years. In the United States, most biodiesel is made from soybean oil or recycled cooking oils. In Tucson, Arizona, for instance, several Mexican restaurants contribute their used vegetable oil to a biodiesel program and even allow individual consumers to bring in used cooking oil to add to the mix.

Still another source of bioenergy is methane recovery. Microorganisms produce biomass gas as they decompose, as do organic materials such as garbage, wood chips, grass clippings, and seaweed. This biomass gas contains methane and carbon dioxide, and it can be collected at landfills and purified to isolate the methane, which can then be used as fuel. Scientists have long been interested in the industrial possibilities of extracting methane from another abundant source—namely, manure, which produces huge amounts of it—and individual energy farmers, so to speak, have been doing so at a smaller scale for several years now, even if the whole biogas enterprise has a Mad Max feel to it.

Agricultural researchers, chemists, and other scientists are actively seeking still more sources of fuel, as well as simpler methods of converting biomass into fuel—a process that is now expensive and complicated. New fuel blends such as biobutanol are more efficient than bioethanol, and some forms of ethanol produced in this country are better than others, but they still need a little more oomph in order to be truly competitive—and to match that Brazilian potion.

Even so, bioenergy is second only to hydropower as the largest source of renewable energy in the world. Biomass is increasingly taking the place of petroleum in the manufacture of goods such as chemicals, paints, and construction materials. It is estimated that by 2010, biomass power will contribute 10,000 megawatts of electrical capacity to the overall supply in the United States, and the federal government recently mandated that ethanol replace 15 percent of projected gasoline use by 2017.

Biodiesel can be made from any fat or oil. Current U.S. biodiesel production is primarily from oil from soybeans such as these, or from recycled restaurant cooking oil. Cleaner burning and renewable biodiesel is most often blended at 20 percent with petroleum diesel. *Left Photo: Warren Gretz, NREL*

Triple biofuels dispenser at Baca Street Biofuels Station in Santa Fe, NM. *Photo: Charles Bensinger and Renewable Energy Partners of New Mexico*

The 2005 Energy Act mandated that 250 million gallons of ethanol be produced from cellulose materials by 2012.

NREL researchers work on the thermochemical pretreatment step of converting lignocellulosic biomass to fuel ethanol and other valuable chemicals.
Photo: Warren Gretz, NREL

Careers in Bioenergy

Bioenergy requires, at the outset, men and women who are skilled at making biomass: farmers, gardeners, foresters, growers of all kinds. It requires people who aren't afraid of getting their hands dirty—or greasy.

Apart from that, the bioenergy field requires a broad range of blue-, white-, and green-collar workers, from sales and support staff to truck drivers and warehouse workers to microbiologists, chemists, and biochemists. Engineers, plumbers, and construction workers are needed to design and build bioenergy plants, while electrical and mechanical technicians, mechanical, electrical, and chemical engineers, mechanics, and heavy-equipment operators are needed to run and maintain them. And researchers are needed to identify and develop new, more efficient sources of bioenergy.

To give you a better idea of that range, the U.S. Department of Agriculture hosts a web site devoted to jobs at the Agricultural Research Service (see the resources section at the end of this chapter). The range of positions listed there shows just how many careers the bioenergy revolution is capable of launching. So, too, does this announcement from a single firm that does bioenergy work in the United States and Europe, listing these job openings in a single week in the fall of 2007:

Mechanical Piping Engineer
Architectural/Civil/Structural Engineer
Aspen Specialist
Business Development Project Manager
Civil/Structural Engineer
Commodity Manager

Cost Estimation Specialist
Electrical Engineer
Engineering & Construction Manager
Ethanol Sales Coordinator
Hydrolysis & Fermentation Engineer
Instrument/Control Engineer
Mechanical Engineer
Operability & Startup Manager
Pretreatment & Fractionation Engineer
Process Engineer
Process Engineering Project Manager
Project Director
Senior Bulk Material Handling Specialist
Senior Process Engineer
Senior Separations Specialist

A reseacher works on cellulose production fermentations to make enzyme for biomass-to-ethanol process.
Photo: Dave Parsons, NREL

Most of those positions are highly specialized, requiring a broad range of skills and training. One of them, that of bioenergy process engineer, for instance, "develops environmentally friendly processes for fractionating plant biomass from crop residues, such as wheat straw, rice straw, rice hulls, and other under-utilized agricultural fibers. The ultimate aim is to produce predictable compositions for use in the creation of composites, nanocomposites, and value-added products, such as specialty chemicals and/or ethanol."

Another position at another firm, that of biomass industry development coordinator, is more green-collar-oriented, involving a blend of technical and sales and communications skills. The coordinator, the firm specifies, "gathers information on competitors, prices, sales, and methods of marketing and distribution to enhance success; using survey results to create a marketing campaign based on regional preferences and buying habits. Business (economic resource) and implementation planning is undertaken by either the individual coordinator or their represented organization to assure viability. The coordinator will solicit business from a variety of prospective commercial customers of varied size and industry and will assist in the assessment of data and methods of service delivery."

Finally, many positions are available at the research level, most requiring specialized training and an advanced degree. Here, for instance, is a listing for entry-level fellowships in bioenergy at Princeton University, which suggests a heady set of skills:

> Post-doctoral positions are available at Princeton University in the area of bioenergy to join a U.S. Department of Energy–funded interdisciplinary team studying microbial hydrogen production from a quantitative perspective. Participating labs include Rabinowitz (metabolomics), Dismukes (photosynthesis), and Rabitz (quantitative modeling). Candidates are preferred with experience in cellular metabolism, microbial physiology, systems biology, mass spectrometry, instrument development, and/or computational chemistry. Exceptional opportunities are available for highly motivated candidates with strong publication records, regardless of their specific area of expertise.

Knowledge Is Power

The Princeton job listed above requires a doctorate, as do many research posts, particularly within the university. The bioenergy process engineer position calls for a recent master's degree in engineering, fiber technology, materials science, or chemistry, adding, "engineering backgrounds may include chemical, mechanical, agricultural, materials, fiber, or process engineering."

McNeil Generating Station in Burlington, Vermont uses wood fuel. *Photo: Dave Parsons, NREL*

The biomass industry development coordinator position calls for an undergraduate degree. But note the last line of the job listing:

> Minimally a Bachelor's degree is required in Business Administration, Economics, or Marketing Research, with a minor in biology, chemistry, engineering or environmental science. Legal studies and law degrees that include a science background may also be good preparation for a career in this field. Basic understanding of scientific concepts and terms is key. Master's and Ph.D. degrees required for advancement.

In short, many of the positions within the field of bioenergy require at least an undergraduate degree, with a marked preference shown for graduate work. Fortunately, the field is broadly defined both within the realms of engineering and agriculture, programs that hundreds of colleges and universities offer. See the listings at the back of the book for some of them, and narrow your search by using College Source Online (*www.collegesource.org*) or a similar web site. ❖

The U.S. uses more than 140 billion gallons of gasoline and almost 40 billion gallons of diesel fuel annually.

More than 60 percent of the petroleum we use is imported.

SOURCE: BIOMASS CONVERSION RESEARCH LABORATORY, MICHIGAN STATE UNIVERSITY

MEDIAN SALARIES/WAGES ❖ Bioenergy		
Agronomist	$35,230	– $59,370
Biochemist	$40,030	– $64,180
Chemist	$35,180	– $60,000
Microbiologist	$41,430	– $66,780
Chemical Engineer	$48,670	– $70,570
Material Scientist	$50,470	– $90,670
Agricultural Scientist	$37,140	– $53,740
Agricultural Engineer	$45,170	– $57,710

SOURCE: WWW.PAYSCALE.COM (DECEMBER 2007)

Making Biodiesel

Michael Guymon is the president of Arizona BioFuels, a start-up company that plans to distribute biodiesel fuel in Arizona (see www.azbiofuels.com). In this conversation with *Tucson Weekly* columnist Jim Nintzel, he looks at some alternatives to ethanol.

What does your company do? *My company will manufacture biodiesel fuel. It doesn't manufacture it yet, but it will. We are talking to a company that will produce oil from algae. This company will cultivate algae and press the algae to extract the oil, and we will purchase the oil and turn it into biodiesel.*

How did you come across this algae thing? *There's a professor at the University of New Hampshire, of all places, who published a report in 2004 that basically said the holy grail of the biodiesel industry is going to be algae, because you can extract 100 to 200 times more oil per acre than soy. So just from an efficiency standpoint, algae make a lot of sense. He also went on to say that the best place to cultivate algae is the Salton Sea, which is between here and San Diego, near Palm Springs. Once I obtained that report, I started looking at building the business around algae oil. At that time, I thought that algae oil was five or ten years away. I've come to find out recently algae oil could be as close as one year away.*

Are we seeing biodiesel catch on with more people? *I see more and more vehicles on the road running on biodiesel. Safeway just announced that all of their fleet will run on 20 percent biodiesel, so there are national companies that are taking the big step that needs to be taken to show that these big rigs can run on biodiesel without any problems. The city of Tucson already has a policy that all of its diesel fleet runs on 20 percent biodiesel.*

Why the 20 percent figure? *Because the original engine manufacturers—car companies, essentially—right now are only comfortable with a 20 percent blend.*

Is it primarily an environmental advantage to switch over, or is it a financial one as well? *Definitely an environmental advantage, because a lot of companies are trying to show how green they are. Even News Corp. recently made the announcement that they were going green. So there are a lot of companies trying to find a way to go green. But financial advantage? It really depends. Right now, biodiesel, even at the wholesale level, is probably selling equivalent to diesel. From a financial standpoint, it's probably a wash.*

The Carbon Cycle of Biofuels

solar energy is absorbed by biomass crops

Sunlight

biomass containing carbon is processed into biofuels

carbon dioxide is absorbed by plants as they grow

carbon dioxide is released as fuel burns

biofuels power your vehicle

SOURCE: *HYDROGEN—HOT STUFF COOL SCIENCE* BOOK

Bioenergy Resources

ORGANIZATIONS & WEB SITES

American Bioenergy Association

ABA advocates for all aspects of biomass development: power, fuels, and biobased chemicals. The trade association does policy work in hopes of getting the government to fund and support development projects. They also provide outreach to farmers and environmental organizations on the benefits of biomass products.

209 Pennsylvania Avenue SE
Washington, DC 20003
(202) 467-6540
www.biomass.org

Biomass Energy Businesses in the World

This web site lists bioenergy firms of various kinds and is a useful first stop in planning a job search.

energy.sourceguides.com/ businesses/byP/biomass/biomass. shtml

University of Illinois Center for Advanced Bioenergy Research

This blog, produced by the Center for Advanced Bioenergy Research (CABER) at the University of Illinois, offers a useful roundup of research news and related topics dealing with biofuels. It is regularly updated.

bioenergyuiuc.blogspot.com/

U.S. Department of Agriculture Careers with the Agricultural Research Service
This federal web site contains listings for a wide variety of positions within the Agricultural Research Service, where significant work in biofuels is being carried out.
www.ars.usda.gov/Careers/Careers.htm

U.S. Department of Energy Energy Efficiency and Renewable Energy: Biomass Program
This web site offers abundant resources for students, including information on various biomass technologies and sites devoted to jobs and careers.
www1.eere.energy.gov/biomass/for_students.html

PUBLICATIONS

Caye Drapcho, John Nghiem, and Terry Walker, *Biofuels Engineering Process Technology* (McGraw-Hill Professional Publishing, 2008)
Written by a team of experts from Clemson University with experts from around the world, this book is a comprehensive description and discussion of the concepts, systems and technology involved in the production of fuels on the industrial and individual scales.

Lyle Estill, *Biodiesel Power* (New Society, 2005)
Lightly touching on the technical aspects of the fuel, its qualities and specifications, this book is largely about the people and stories of the biodiesel movement.

William Kemp, *Biodiesel, Basics And Beyond: A Comprehensive Guide to Production And Use for the Home And Farm* (Aztext Publishing, 2006)
A useful handbook on how to make biodiesel fuel, including processes, lists of equipment, and a compendium of reference materials and suppliers.

Lisbeth Olsson, ed., *Biofuels* (Springer, 2007)
A Ph.D.-level series of technical papers on various topics in bioenergy, useful for giving a sense of present and future research emphases.

Greg Pahl, *Biodiesel: Growing a New Energy Economy* (Chelsea Green, 2005)
Written for a nontechnical readership, Pahl's book surveys the history of the biofuels industry, assessing successes and current shortcomings and offering insights into its future.

Josh Tickell, *Biodiesel America* (Yorkshire Press, 2006)
Tickell shows that an abundance of available, economically viable, and profitable energy solutions exists, and at the forefront of these new energy technologies is biodiesel.

Hydrogen Energy & Fuel Cells

6 Hydrogen Energy & Fuel Cells

Hydrogen, one of the building blocks of the universe, is abundant. It also burns very cleanly, with the sole byproduct of burning being—well, water, for which reason talk of a "hydrogen economy" has been with us for many years. That economy is some distance off: hydrogen is difficult to transport, since it is highly flammable and requires considerable pressurization, and mass-production facilities have yet to be built. Still, a hydrogen-based fuel cell is around 60 percent efficient at converting fuel to power, double the efficiency of an internal combustion engine, and efficient compared to many other kinds of renewable energy storage and delivery systems. At some point in the not too distant future, hydrogen is going to play an important part in the world's energy mix, and making that happen will open up many opportunities for scientists, technicians, and other workers.

Buses are just one of the many places that fuel cells are used. *Photo: Hydrogenics (www.hydrogenics.com)*

One man who has been working in the field for many years now is Giorgio Zoia, who saw his first hydrogen-powered car at an exhibit

5 Main Types of Fuel Cells

◆ alkaline
◆ molten carbonate
◆ phosphoric acid
◆ proton exchange membrane
◆ solid oxide

A fleet of 100 fuel cell-powered Chevy Equinoxes will be tested in 2007. The chassis illustration (right) shows three carbon-fiber fuel tanks, lithium ion battery pack, and the fuel cell stack.
Photos © General Motors

in his native Italy. He was so impressed by it that he set out a rigorous course of education in science and mechanics, and for many years he has worked in the renewable-energy division at British Petrol (BP). Recently, he collaborated with DaimlerChrysler to maintain a fleet of 30 of the approximately 100 fuel-cell cars in California, where BP has its research facility. Each of those cars costs a million dollars. But, Zoia says, their price will fall as hydrogen production and fueling stations come online—perhaps as soon as 2012.

One promising area of research is the use of photochemical molecular devices to produce hydrogen gas from water. Chemists from Virginia Tech unveiled devices called supramolecular ruthenium(II), rhodium(III) mixed metal complexes that use the energy from light to collect electrons, which are then used to split the hydrogen and oxygen in water. The researchers have been striving to make the process more efficient and in the past year, they have come up with additional molecular assemblies that absorb light more efficiently and activate conversion more efficiently. "We have come up with other systems to convert light energy to hydrogen," says Professor Karen Brewer. "So we have a better understanding of what parts and properties are key to having a molecular system work. Previously we concentrated on collecting light and delivering it to the catalyst site. Now we are concentrating on using this activated catalyst to convert water to hydrogen. Once we know more about how this process happens, using our supramolecular design process, we can plug in different pieces to make it function better."

An existing technology that is widely employed in Britain is the combined heat and power (CHP) system, which produces both electricity and usable heat energy. One, used in the town hall of Woking, England, has a fuel cell that uses phosphoric acid as the electrolyte, natural gas reformed into hydrogen as the fuel, and oxygen taken directly from the air as the oxidant. The fuel cell is able to supply 200kW of electrical power. Future developments will see the fuel cell operating alongside other renewable energy sources such as solar power to provide a greater capacity.

In Britain as well, bioscientists at the University of Birmingham recently demonstrated a reactor that uses hydrogen-producing bacteria and sugar waste to generate electricity via a fuel cell. Interestingly, the reactor uses waste products such as diluted nougat and caramel waste from the candy industry as food for the bacteria, waste that usually ends up in landfills. Once in the reactor, the bacteria consumed the sugar, producing hydrogen and organic acids. A second type of bacteria was then introduced into a second reactor to convert the organic acids into more hydrogen. The hydrogen was then fed to a fuel cell, in which it reacted with oxygen in the air to generate electricity.

Honda solar-powered water electrolyzing hydrogen station. *Photo: Honda*

Ways to Liberate Hydrogen

- **Electrolysis** split water with electricity
- **Steam Electrolysis** split water with heat, pressure and electricity
- **Direct Solar Thermal Water Splitting** split water with heat
- **Photoelectrochemical** split water using sunlight directly
- **Thermochemical** split water using chemicals and heat
- **Biological** split water using microorganisms
- **Steam Reforming** convert the methane in natural gas using steam
- **Direct Thermal Splitting of Natural Gas** split natural gas using heat
- **Gasification** breakdown coal or biomass with heat and pressure

SOURCE: *HYDROGEN—HOT STUFF COOL SCIENCE* BOOK

In the bargain, waste biomass left behind by the process was removed, coated with palladium, and used as a catalyst in another process to remove pollutants such as chromium and polychlorinated biphenyls (PCBs) from the environment.

Careers in Hydrogen Energy and Fuel Cells

Although the hydrogen economy lies in the future, many opportunities exist today. Many are in the research field, largely within academia or the government, while other technical positions are available at private firms.

A technician builds a Fuel Cell Power Module in the Hydrogenics plant.
Photo: Hydrogenics (www.hydrogenics.com)

One solid-oxide fuel-cell (SOFC) manufacturer, for instance, recently advertised for a manufacturing equipment engineer. The job requires a bachelor's degree in mechanical, electrical, or ceramic engineering or equivalent experience; two to eight years of related field experience; general knowledge of manufacturing processes; experience with structured problem-solving processes and process-capability techniques; knowledge of PFMEA (process failure modes and effects analysis), equipment design, ergonomics, and safety standards as related to manufacturing processes.

Another manufacturer posted a listing for a process manager in the company's research and development division. The specific job responsibilities included:

◆ Developing innovative and enabling SOFC materials, process, and component technologies.

◆ Writing successful research proposals to private and governmental research sources.

◆ Establishing and directing experimental research plans for multiyear research programs.

- Contributing to the company's intellectual property portfolio.
- Assuring the transition of technologies from laboratory to commercial products.
- Managing costs and resource utilization while achieving project goals.
- Producing comprehensive reports detailing technical progress and plans.
- Presenting research results at technical conferences.

Critical qualifications, the company added, included a PhD in materials science, chemical engineering, or a related discipline, and a minimum of five years of technology development experience in fuel cells or other high-temperature ceramic electrochemical systems. "The ideal candidate will have a comprehensive understanding of the state-of-the-art in fuel cell technology, a worldwide reputation for technical achievement and innovation, and a proven track record for solving problems and achieving technical success," the listing concluded. "Prior technology commercialization experience is of particular interest."

And Ford Motor Company recently advertised several openings for fuel-cell engineers, encouraging applications from candidates with a strong background in any of the following fields:

- Fuel cell performance and life/degradation functions, modeling and model validation.
- Fuel cell design, development and validation.
- Fuel cell materials development, including analysis and qualification.
- Fuel cell reliability/durability, focused on detailed understanding of physical mechanism of failures.

The U.S. Navy has used fuel cells in submarines since the 1980s, and alkaline fuel cells have flown over 100 missions and operated for over 80,000 hours in spacecraft operated by NASA. *Photo: NASA*

◆ Fuel cell manufacturing and cost analysis.

◆ Fuel cell, MEA and materials testing and hydrogen fuel cell laboratory operation.

Candidates for these positions, Ford added, should have the following educational and vocational experiences and skills:

◆ M.S. or higher degree in Electrochemical, Electrical, Mechanical or Chemical Engineering with a specialty in fuel cell research and development or fuel cell computational modeling.

◆ At least three years of hands-on experience developing or testing fuel cells and fuel cell materials.

◆ Excellent teamwork and communication skills.

◆ Self-motivated with the ability to manage multiple tasks.

◆ Proficient in computer skills.

An NREL researcher works on the compressor of the Wind to Hydrogen Project (Wind2H2). *Photo: Ben Kroposki, NREL*

For more on careers in hydrogen energy and fuel cells, visit the job bulletin board at Fuel Cell 2000 (*www.fuelcells.org/ced/career/fcjobpages.htm*). The work is there—so hit the books!

Knowledge Is Power

Creating the hydrogen economy will require the skills of many researchers and technicians, skills that are acquired almost entirely in universities, technical schools, and laboratories. Fortunately, the federal government and other agencies have been generous in funding scholarly research, and almost every state has at least one center for hydrogen fuel-cell studies.

Hocking College, for instance, a two-year institution in central Ohio, offers an associate's degree in fuel-cell technologies. Dean Jerry Hutton expects enrollment to grow rapidly as students learn more about alternative-energy careers. "It's time to bring it home," he tells the *Columbus Dispatch*. "Ohio, in the industrial revolution, we were very much there. We've given a lot of those jobs away. We need to be in that global market. It's a green revolution. We're creating green-collar jobs, instead of white- or blue-collar."

Some of the most prominent or active are listed below; for a complete list, visit the U.S. Department of Energy's Energy Efficiency and Renewable Energy web site (*www.eere.energy.gov/hydrogenandfuelcells/resources/db/higheredcatalogs*), a comprehensive roster of every school now engaged in such work. ❖

Iceland has started to convert its fishing fleet from diesel engines to hydrogen fuel cells as part of a national project to create a fossil fuel free economy.

Alfred University
Center for Environmental and Energy Research (CEER)
2 Pine St., Alfred, NY 14802
ceer.alfred.edu

Arizona State University
Center for Applied Nanobioscience
1001 S. McAllister Ave., Tempe, AZ 85287
www.biodesign.asu.edu/centers/anb/overview

California Institute of Technology
Department of Material Science
1200 East California Blvd.
Pasadena, CA 91125
www.matsci.caltech.edu

Colorado School of Mines
Colorado Fuel Cell Center
1310 Maple Street, Golden, CO 80401
www.coloradofuelcellcenter.org

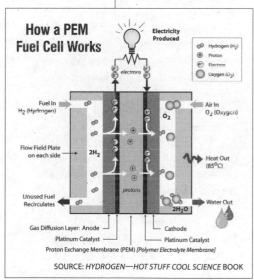

How a PEM Fuel Cell Works

SOURCE: *HYDROGEN—HOT STUFF COOL SCIENCE* BOOK

Columbia University
Earth Engineering Center
New York, NY 10027
www.seas.columbia.edu/earth/index.html

Cornell University
Cornell Fuel Cell Institute
330 Bard Hall, Ithaca, NY 14853
www.cfci.ccmr.cornell.edu

Fresno City College
Advanced Transportation Technology
1101 E. University Avenue, Fresno, CA 93741
www.fresnocitycollege.edu/appliedtech/att/

Illinois Institute of Technology
Center for Electrochemical Science and
Engineering
10 W. 33rd Street, Chicago, IL 60616
www.chbe.iit.edu/research/cese/cese.html

Pittsburg State University
Dept. of Engineering Technology, Room W223
Pittsburg, KS 66762
www.pittstate.edu/etech/

Princeton University
Energy Group
Guyot Hall, Washington Rd., Princeton, NJ 08544
www.princeton.edu/~energy/index.html

Tennessee Technological University
Center for Manufacturing Research
P.O. Box 5077, Cookeville, Tennessee 38505
www.tntech.edu/cmr/

Focus on Fuel Cell Education

Fuel cells involve many branches of science and many technologies, and understanding them therefore requires solid grounding in several disciplines.

At the Colorado Fuel Cell Center, a division of the renowned Colorado School of Mines, students must master the basics of chemical engineering, electrical engineering, mechanical engineering, and materials science. A specialized course follows in fuel-cell science and technology, introducing students to the fundamental aspects of fuel-cell systems, with emphasis placed on proton exchange membrane (PEM) and solid oxide fuel cells (SOFC).

Students go on to learn the basic principles of electrochemical energy conversion, relevant topics in materials science, thermodynamics, fluid mechanics, electrochemistry, fuel-cell operation, system integration, and fuel processing. That's a curriculum!

Texas State Technical College
Fuel Cell Technology
3801 Campus Drive, Waco, TX 76705
www.waco.tstc.edu/ecr/ecr_fuelcell/index.php

University of California, Davis
Hydrogen Production and Utilization Lab
1236 Bainer Hall, Davis, CA 95616
mae.ucdavis.edu/hypaul/

University of Florida
Fuel Cell Research and Training Laboratory
SW 23rd Terrace, Gainesville, FL 32610
grove.ufl.edu/~fuelcell/

University of Minnesota
Initiative for Renewable Energy
and the Environment
1500 Gortner Avenue, St. Paul, MN 55108
www1.umn.edu/iree/hydrogen.html

Virginia Polytechnic Institute and State University
Center for Automotive Fuel Cell Systems
100S Randolph Hall, Blacksburg, VA 24061
www.me.vt.edu/research/centers/afcs.html

MEDIAN SALARIES/WAGES ❖		
Hydrogen Energy & Fuel Cells		
Chemical Engineer	$48,670	– $70,570
Material Scientist	$50,470	– $90,670
Material Engineer	$54,980	– $90,219
Mechanical Engineer	$52,230	– $85,320
Research Scientist	$51,320	– $99,650
SOURCE: WWW.PAYSCALE.COM (DECEMBER 2007)		

The BMW 745h V-12 is being developed with a
hydrogen-fueled internal combustion engine
instead of fuel cells. As a dual-fuel vehicle, it has
two fuel tanks, one for liquid hydrogen and one
for gasoline. *Photo © BMW (GB) Ltd.*

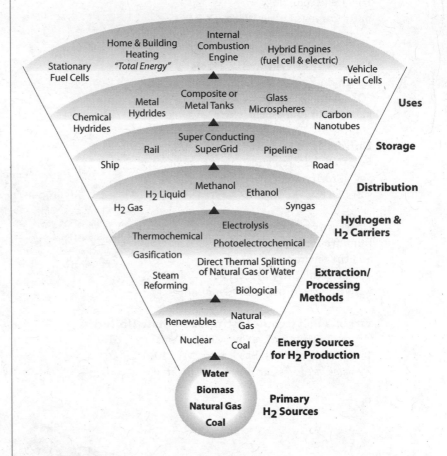

Moving Towards a Hydrogen Economy

Numerous job opportunities are available in all of the layers shown above, from hydrogen extraction and processing to distribution, storage, and applications.

SOURCE: *HYDROGEN—HOT STUFF COOL SCIENCE* BOOK

Hydrogen Resources

ORGANIZATIONS & WEB SITES

4Hydrogen
Check out their Hydrogen Directory for a good list of links to government, non-profits, universities, infonets and corporations. They also have a good introduction to hydrogen as fuel.

www.4hydrogen.com

Florida Solar Energy Center
In 1997, FSEC was designated a Center of Excellence in Hydrogen Research and Education by the U.S. Department of Energy. The development of applications and industries that use FSEC's hydrogen and fuel cell research results has been a primary goal.

1679 Clearlake Road
Cocoa, FL 32922-5703
(321) 638-1000
www.fsec.ucf.edu

Fuel Cell Today
A London-based organization offers market-based intelligence on the fuel cell industry. They offer an online Career section, categorized by Academic, commercial & sales, Engineering, Managerial, Research, and Other.

www.fuelcelltoday.com

International Association for Hydrogen Energy
The goal of this IAHE is to stimulate the exchange of information in the hydrogen energy field through its publications and sponsorship of int'l workshops, short courses and conferences, and they endeavor to inform the general public about the role of H2 energy. Their official scientific journal, *International Journal of Hydrogen Energy*, is published monthly. The web site provides recent international news, numerous links, conference details, and more.

www.iahe.org

National Fuel Cell Research Center
An excellent site for fuel cell information, activities, tutorials, FAQs, info about the University of California, Irvine's academic program, plus upcoming events, industry publications, educational resources and much more.

www.nfcrc.uci.edu

National Hydrogen Association
Since 1989, NHA has provided data and educational materials to the media, safety and codes & standards officials, policy-makers, and the general public. They have over 100 members, including major industry, small business, government and university organization. Read the latest hydrogen headlines online, and check out their Hydrogen and Fuel Cell Job Board.

www.hydrogenassociation.org

National Renewable Energy Laboratory

Their web site provides a wealth of information about hydrogen and fuel cells, including their R&D efforts on H2 production and delivery, storage, fuell cells, technology validation, safety, codes and standards, and analysis. Learn how research activities crosscut and contribute to advances across the laboratory—in photovoltaics, bioenergy, transportation, wind, buildings, and basic sciences.

www.nrel.gov/hydrogen

U.S. Fuel Cell Council

An industry association whose members include the world's leading fuel cell developers, manufacturers, suppliers and customers. Their site provides an excellent list of books, studies and reports, online publications, plus access to their monthly *Fuel Cell Connection* newsletter, and quarterly *Fuel Cell Catalyst* newsletter.

www.usfcc.com

PUBLICATIONS

Fuel Cell Magazine

A bimonthly publication that serves managers and technical professionals involved in developing and applying fuel cell technologies worldwide. Subscriptions are free to qualified recipients.

www.fuelcell-magazine.com

Gregor Hoogers, **Fuel Cell Technology Handbook** (CRC, 2002) A comprehensive book about the principles of fuel cell technology, plus various fuel cell applications and an expert's look at future developments.

Hydrogen & Fuel Cell Letter

Since 1986 editor Peter Hoffman has covered the science, business, economics, and politics of hydrogen and fuel cells—nationally and internationally. Published monthly.

www.hfcletter.com

Rex A. Ewing, **HYDROGEN—Hot Stuff Cool Science 2nd Edition: Discover the Future of Energy** (PixyJack Press, 2006) An enjoyable read for all ages, and a super resource for anyone looking to sort out the basics of hydrogen science.

Peter Hoffman, **Tomorrow's Energy: Hydrogen, Fuel Cells, and the Prospects for a Cleaner Planet** (MIT Press, 2001) A review said, "this book has everything the reader needs to know about hydrogen—its discovery, the numerous attempts to use it as a fuel, its (quite good) safety record, and the practical and economic difficulties that must be overcome if hydrogen is to realize its potential as a nonpolluting, non-carbon-emitting fuel."

Green Building

7 Green Building

The automobiles we drive and the airplanes that we fly use a lot of energy. The buildings in which we live, work, study, and shop are major consumers of energy, too: according to the U.S. Environmental Protection Agency, buildings account for 39 percent of total energy use, 12 percent of total water consumption, 68 percent of total electricity consumption, and 38 percent of the carbon dioxide emissions in the nation. As scholars, scientists, inventors, entrepreneurs, and workers of all kinds have turned their attention to improving the efficiency of vehicles, so they have increasingly studied how the built environment can be bettered.

NREL's cutting-edge Science & Technology Facility, completed in 2006, has received a LEED Platinum rating. *Photos: Bill Timmerman (above); Marjorie Schott (left), NREL*

And how it can! Look in your own home. Forget for a moment about the appliances and electricity-using things within it; look instead at its veins and bones and skin and eyes, at its pipes and frame and covering and windows.

If there's a dripping faucet, it's dripping at a rate of as much as 20 gallons a day, a drop at a time.

If the house is drafty, then it can lose the same amount of heat or cooled air through leakage as it would if a window-sized hole were simply punched through a wall and left open to the elements.

This off-grid mountain home was designed to make the most of the sun's energy by using passive solar design, a Trombe wall, and photovoltaics. The south-facing sunspace (shown below) grows tropical plants year-round, even at 9,100 feet. *Photos: LaVonne Ewing*

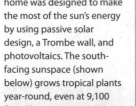

The loss from little leaks can amount to more than a third of your annual utility bill.

If the roof is old and in poor repair, then cool air is fleeing in summer and hot air is running away in winter. An old roof can add 10 or 15 percent to the base cost of your utility bill, and even more.

If the windows rattle, then they're leaking, too. Old windows will do that. A single-glazed, one-pane window has a thermal resistance value, or R-value, of 1: the window is almost the same temperature as the outdoors. Double-glazed windows, with an R-value of 2, are more efficient. "Superwindows," with panes separated by argon or krypton gas or some other insulating substance, have an R-value of between 6 and 10.

Builders have known these realities for a very long time, but it's only been in recent years that technologies have been available to take full advantage of a building's ability to provide shelter. The quest to renovate, and rebuild, and build from scratch intelligently, economically, and sustainably falls under the informal rubric "green building."

There's green in more ways than just environmental friendliness and energy efficiency. Ten years ago, the phrase "green building" might have meant something to a handful of theorists and the visionaries of Berkeley and Cambridge. Today, the green building industry is worth $12 billion, and the numbers are climbing by leaps and bounds.

"If you're going to be part of the construction business for the next thirty or forty years, then you're going to have to make sustainability and green building a big part of your repertoire," says Gregory Esau, a carpenter and builder in Vancouver, Canada. "A lot of municipalities, for instance, are going carbon-neutral, and we're going to have challenges figuring out how to do that—not just in the buildings themselves, but also in building the buildings, what with all the big machines we have to use! We'll need good people to help us figure out solutions to problems like that, workers who are on what I call a green track."

Asked for advice he might give to a student with an interest in green building, Esau enthusiastically counsels, "Try to be a leader! Do your homework, learn as much as you can, ask a lot of questions. You can make a good living at this work, and it's a very exciting time to get into it."

Getting Into Green Building

Scott Sklar, the energy-market specialist whom we met in the introduction, writes regularly for several online publications in the field. Here's his answer to an inquiry from a resident of upstate New York who asked about where a person interested in a career in designing residential and commercial projects might start.

I am receiving many e-mails from Renewable EnergyAccess.com readers on jobs in the industry, particularly in the design, sales, installation and service sector of the industries.

First, while the clean energy industries are growing more than 30 percent per year, the biggest bottleneck in this growth is in the product delivery chain to the customer. When I get asked, "How should I enter the field?" my first response is to approach existing service vendors.

In system design, architectural and engineering firms are a good start in your locality. They may have someone, either full or part time, assigned to the renewable or distributed energy or green building sectors, and they may want to add or train someone in providing services. So check with people locally or go to the local Yellow Pages.

Second, the existing solar integrator, design, sales, installations and maintenance firms are growing, and they may need to add employees. Contact them, again checking the Yellow Pages in your area, or even better at FindSolar.com or via the State Chapter lists from the websites of the American Solar Energy Society or the Solar Energy Industries Association.

I also urge approaching corollary industries that might want to set up some kind of parallel business in addition to their core business: HVAC contractors, building security companies, local telecommunications and cellular/tower companies, and traditional backup power companies now primarily offering diesel or battery banks—since they may want to expand to cleaner applications. (RenewableEnergyAccess.com lists job opportunities, which can be broken down to suit your search.)

Local unions, such as IBEW, have a concerted effort to train their members, and I am sure local electricians, carpenters and builders are also beginning to add jobs in this field. So think out what may be best in your area, reach out to established and new players.

Good luck in finding a job. We need lots of new blood in this field as we scale up and grow.

Careers in Green Building

Green building, one source defines it, is "the practice of creating healthier and more resource-efficient models of construction, renovation, operation, maintenance and demolition." That's a tall order, particularly when it comes to the challenge of refitting old buildings greenly (so to speak) and taking down the ones that cannot.

A green building requires visionaries. It requires women and men who can study the land to determine where the sun will fall in winter and summer and where the wind will blow, the better to create efficient passive-solar systems for heating and cooling, the better to place energy-generating photovoltaic panels and wind turbines, the better to know where a geothermal pipe will find its happiest home. It requires architects, contractors, and builders who are committed to finding the most sustainable path to the best building they can make. It requires real-estate agents, lawyers, salespeople, accountants. It requires scientists who understand the physics of heat transfer, the flow of water, the movement of the winds. It requires engineers of every description. It requires men and women who can put those systems into place and maintain them: carpenters, plumbers, electricians, roofers, landscapers, glaziers and window installers. The list goes on and on.

In short, there's room for just about anyone in the green building field—anyone, that is, who is willing to work hard, learn, and keep working hard and learning for all the years to come.

The Florida Solar Energy Center uses solar hot water, solar electricity, low-e windows, and light "shelves" for reflected light , just to name a few of its energy efficient features. *Photo: Nicholas Waters, FSEC*

Knowledge Is Power

Green building, at whatever level and whether approached

from the white-collar, blue-collar, or green-collar perspective, requires specialized knowledge. Workers in the building trades are often expected to have that training in hand before going on the job, acquired from community-college or trade-school courses and certifications. One interesting sequence of courses, apparently geared to workers with on-the-job skills, is Mount Hood Community College's Sustainable Building Advisor Training Program (*www.mhcc.edu/sbap*).

For training in the trades in your area, do an Internet search combining the terms "green building" and "community college," along with your location. Doing so in the city in which I live turns up, for instance, a site from Pima Community College devoted to Construction Resources, and it's well worth a look wherever you are: the site is at *www.pima.edu/ library/online-distance/study-guides/Construction.shtml*.

Workshops such as those offered by Solar Energy International (*www.solarenergy.org*) and Yestermorrow (*www.yestermorrow.org*) can provide basic training and continuing education, and apprenticeships are opening up as more and more skilled workers arrive on the scene and are available to serve as mentors.

For most "green-collar" positions, a college degree is preferred. Here, for instance, is a listing for a position as a green building project analyst working in the Southeast:

> Provide sustainable design support services for planning, implementing, and follow-up to ensure the facility's construction projects fully meet the goals of sustainable design. Participate in early construction project planning. Attend, coordinate, and facilitate (senior level) LEED® [Leadership in Energy and Environmental Design] credit dialogue and negotiations during planning and design charrettes; and assist client staff, as needed, in managing the sustainable

Natural daylighting is used at NREL's Solar Energy Research Facility. *Photo: Warren Gretz, NREL*

Fenestration Systems
Designing, orienting, shading and sizing of windows to increase their benefits while reducing problems (such as glare, overheating, condensation, large heating/cooling loads) is an important goal of the building designer.
SOURCE: FLORIDA SOLAR ENERGY CENTER

Hands-on workshops in natural home building are a good place to start learning. *Photo: A Strawbale Construction course at SEI. (www.solarenergy.org)*

design portion of each project. Provide technical guidance/ recommendations for potential LEED credits that may be achievable and practical. Provide assistance in adding credits to each project's sustainable design rating whenever feasible. Identify initial sustainable design credit goals and identify any changes to those goals as they occur and include the rationale for the changes. Inform management of any new sustainable design policies or requirements.

The job calls for specialized training—but, at the outset, a college degree:

Minimum Qualifications: Bachelor's degree (preferred Master's Degree) in architecture, civil or environmental engineering, environmental studies, environmental sciences, renewable energy, sustainability planning, natural resources management, or related field. Professional training and work experience in energy efficiency, sustainability planning, master planning and/or sustainable design and construction. Knowledge of, and experience implementing, the U.S. Green Building Council's LEED guidelines; and Federal environmental laws and regulations and pollution prevention programs. Good computer skills in MS Office software are essential. Must have strong organizational and interpersonal skills to successfully coordinate and facilitate sustainable design portions of planning and design charrettes, briefings, and related activities. The on-site personnel must have the ability to work well as a team member.

MEDIAN SALARIES/WAGES ❖ Green Building		
Architect	$51,980	– $63,600
Civil Engineer	$44,010	– $78,400
Electrical Engineer	$55,040	– $96,410
Mechanical Engineer	$52,230	– $85,320
Construction Manager	$65,730	– $90,610
Construction Superintendent (Commercial)	$56,050	– $78,670
Architectural Drafter	$15.10 – $22.40 / hour	
HVAC Service Technician	$15.10 – $22.70 / hour	

SOURCE: WWW.PAYSCALE.COM (DECEMBER 2007)

Taking Efficiency to a New Level

Michael Sykes invented the award-winning Enertia® Building System (U.S. Patent No. 6,933,016) in the mid-1980s *(www.enertia.com)*. As this conversation shows, Sykes affords a textbook example of how technological knowledge and entrepreneurship can lead to a highly successful career in renewable energy and green building.

What motivated you to start building Enertia Homes? *Over the years I had made observations that struck me as out of the ordinary. Then one day they all came together, there was a light-bulb moment and I realized I had stumbled on a new form of energy. I made up a word for it…"Enertia." It would enable us to use nature's most abundant renewable material, wood, to build houses that heat and cool themselves without fuel or electricity. Amazingly, in 1984, no manufacturer or investor was interested so I reluctantly set about to make them myself. I could not let it go—here was a solution to a major worldwide problem. Finding investors and growing the business continues to be my biggest challenge.*

What education and skills do you use as an inventor of efficient green homes? *Enertia, energy from a change in phase, was so new that I had to write my own equations for it. I actually used the math I learned in high school and college, including differential equations. The air-flow patterns are based on physics and meteorology. The materials are based on biology and forestry, using nanotechnology and chemistry. The designs are based on architecture; the structure is based on engineering. To make the houses I had to learn industrial engineering and machinery design. Yet everything I learned as I developed my invention went against the conventional thinking in every field. When your invention is new you have to be an entrepreneur—by definition, no one else is doing it.*

Winter

To achieve super efficiencies, an Enertia home combines the effects of several principles, including the greenhouse effect, the thermal inertia (mass) of wood and earth, and the natural convective movement of heated air. The heavily windowed sunspace—a 4-foot-wide room running along the entire south side of the main level—captures winter heat and transfers it throughout the house via airflow channels between the ceiling and the roof, and the inner and outer north walls. In summer, cool air is pulled from the north side and exits through roof vents. It is, essentially, a solid wood house within an envelope of glass and wood. *www.enertia.com*

Summer

Thousands of colleges and universities offer programs that might lead to such a job. Some are well-established flagship schools such as MIT and Yale, whose architectural programs are second to none; others are more experimental, such as the intriguing, unashamedly hippie New College of California, whose North Bay Campus offers a master's degree in Eco-Dwelling, the coursework for which includes these classes:

◆ Natural Systems, Human Systems
◆ History of Energy and Dwelling
◆ Causal Dynamics
◆ Design with Site and Climate
◆ Permaculture
◆ Small-Space Design
◆ Working with Clients
◆ Natural Heating and Cooling

◆ Water and Waste Systems
◆ Structural Systems
◆ Renewable Energy Systems
◆ Human Environmental Health
◆ Green Building Materials
◆ Building Codes and Permitting
◆ Access to Land
◆ Natural Remodeling

Arcosanti, an experimental community in the Arizona desert. *Photo: Gregory McNamee*

You stand to learn a lot wherever you choose to study, for green building is a field that's being defined and redefined every day, where experimentation and entrepreneurship are at a premium. To find the school that's right for you, ask yourself what kind of work you'd like to be doing ten years from now. If it's building large, self-contained structures and experimental communities such as Arcosanti, in the Arizona desert, then a blend of architecture and engineering such as MIT offers might be just right. If it's siting buildings and communities to take best advantage of the natural environment, then a degree in environmental studies such as the excellent one the University of Washington offers could be the key. Talk to architects, builders, and engineers in your community and take their advice. A green track awaits. ❖

Green Building Resources

ORGANIZATIONS & WEB SITES

BuildingGreen
This company provides accurate and timely information to building-industry professionals and policy makers to improve the environmental performance and reduce the adverse impacts of buildings. Also publishes *Environmental Building News*.
www.buildinggreen.com

Florida Solar Energy Center
FSEC building science program researches and develops building improvement strategies that reduce energy use, enhance the economy, and improve the environment. Research projects include Zero Energy buildings, fenestration, energy efficient schools and green standards.
1679 Clearlake Road
Cocoa, FL 32922-5703
(321) 638-1000
www.fsec.ucf.edu

Green Home Building
A useful web site devoted to sustainable architecture, solar home heating, and hybrid architecture; the listings of workshops alone make the site worth bookmarking.
www.greenhomebuilding.com

Solar Energy International
For workshops on sustainable home design and natural house building, check their website for the latest schedule.
www.solarenergy.org

Sustainable Buildings Industry Council
SBIC was founded in 1980 as the Passive Solar Industries Council. While it continues to be an association of associations committed to high-performance design and construction, its new name reflects its efforts in the fields of architecture, engineering, building systems and materials, product manufacturing, energy analysis, and "whole building" design.
www.SBICouncil.org

U.S. Environmental Protection Agency
The EPA web site provides resources and publications for the study of green building, as well as listings of grants and other funding opportunities for green-building projects.
www.epa.gov/opptintr/greenbuilding

U.S. Green Building Council

With more than 11,000 member organizations and a network of 75 regional chapters, USGBC is a nonprofit organization composed of leaders from the building industry working to promote buildings that are environmentally responsible, profitable and healthy places to live and work. Its web site is a mine of information and resources, including LEED (Leadership in Energy and Environmental Design).

www.usgbc.org

Yestermorrow Design/Build School

They offer over 100 hands-on courses per year in design, construction, woodworking, and architectural craft and offer a variety of courses concentrating in sustainable design.

www.yestermorrow.org

PUBLICATIONS

Chris Magwood, Peter Mack, and Tina Theirren, *More Straw Bale Building: A Complete Guide To Designing And Building With Straw* (New Society Publishers, 2005)
A practical book that covers the entire process of building a bale structure: developing sound building plans; roofing; electrical, plumbing, and heating systems; building code compliance; and special concerns for builders in northern climates.

Ian McHarg, *Design with Nature* (Wiley, 1995)
Blending philosophy and science, McHarg shows how humans can copy nature's examples to design and build better structures.

Sustainable Buildings Industries Council, *Green Building Guidelines: Meeting the Demand for Low-Energy, Resource-Efficient Homes* (Fifth edition, 2007)
A builder-friendly manual that is applicable to homebuilders interested in exploring the notion of rethinking some of the design issues in their current product.

Carol Venolia & Kelly Lerner, *Natural Remodeling for the Not-So-Green-House* (Lark Books, 2006)
Packed with information and color photos, this book is filled with motivational case studies and informative graphics.

Alex Wilson and Mark Piepkorn, *Green Building Products: The GreenSpec Guide to Residential Building Materials* (New Society Publishers, 2006)
You'll find descriptions and manufacturer contact information for more than 1,400 environmentally preferable products and materials in this book. All phases of residential construction, from sitework to flooring to renewable energy, are covered.

Alex Wilson, *Your Green Home* (New Society Publishers, 2006)
Written for homeowners planning a new home, this book sets out to answer some of the big-picture questions relating to having a home designed and built.

Energy Management

8 Energy Management

Buildings are major polluters, and they account for about four-tenths of all the energy we use in this country. On the positive side, though, they also afford an arena in which we can improve our habits of using energy, and building environmentally friendly structures, as we have seen, is one important path toward lessening our energy consumption overall. So is designing more efficient energy systems, putting them in place, and maintaining them.

The field of energy management is a broad one, and it requires familiarity not only with energy technologies but also with the building trades, making many jobs within it an interesting blend of blue- and green-collar work.

An energy engineer, for example, works with computer software to develop various scenarios of energy use for new building projects, then provides hands-on, quite specific advice to owners and contractors on how to improve energy efficiency. Many energy engineers come from the building trades, with backgrounds in HVAC, electrical systems, and general contracting; others have degrees in engineering or architecture and develop such skills by spending time apprenticing on construction sites and in other on-the-job situations.

An energy analyst has some knowledge of engineering, but

A Building America staff member tests a newly built home for energy efficiency by conducting energy monitoring, at the furnace and at the electrical meter. *Photo: Warren Gretz, NREL*

spends most of his or her time on the job performing calculations and audits based on energy use and cost, usually after clients have requested help in identifying ways in which energy costs can be lowered in existing structures. Energy analysts are thoroughly familiar with building codes related to energy. Some analysts come from the building trades, although many energy analysts have training as electrical engineers. Certification is desirable, which requires passing a standards exam and attending yearly training classes and workshops.

HVAC technicians are also essential to the practical work of maintaining energy-efficient heating and cooling systems. In the past, most such technicians came up through the building trades, but because those systems are now so complex, today most come into the field following an apprenticeship or, more often, technical or trade school, with training in electronics and basic math emphasized.

An emerging option is in what Lane Community College, which

A Zero Energy Home in Lakeland, Florida, with solar panels on the roof. *Photo: Steven C. Spencer, FSEC*

Saving Energy

Energy-efficiency and renewable-energy authority Amory Lovins, the head of the Rocky Mountain Institute, has been working to develop "soft-path" solutions to the energy problem since the 1970s. He believes that one key to saving energy is to put better lighting systems into use: a restaurant, he calculates by way of example, can cut 80 percent of its electricity costs simply by installing energy-efficient light bulbs. "There's upwards of a hundred giant power plants to be saved by proper lighting systems," he told writer Elizabeth Kolbert in a *New Yorker* profile published in January 2007.

Another smart solution is to replace old windows with new double-paned, gas-sealed ones; in one Chicago office building, he determined, the savings from that retrofitting—an expensive process—would be so quickly realized that payback would come "in minus five months."

Making calculations of this sort is the job of energy gurus—and energy managers.

we visited in the chapter on solar energy, calls an "energy management technician." Its two-year program is both rigorous and comprehensive, and it gives a good idea of what an interested student might look for in other programs around the country. As the LCC catalog describes it, the graduate will:

◆ Evaluate the energy use patterns for residential and commercial buildings and recommend energy efficiency and alternative energy solutions for high-energy consuming buildings.

◆ Understand the interaction between energy consuming building systems and make recommendations based on that understanding.

◆ Construct energy evaluation technical reports and make presentations for potential project implementation.

◆ Use appropriate library and information resources to research professional issues and support lifelong learning.

◆ Access library, computing and communications services, and obtain information and data from regional, national and international networks.

◆ Collect and display data as lists, tables and plots using appropriate technology (e.g., graphing calculators, computer software).

◆ Develop and evaluate inferences and predictions that are based on data.

◆ Determine an appropriate scale for representing an object in a scale drawing.

◆ Interpret the concepts of a problem-solving task, and translate them into mathematics.

How Much Do Energy Managers Earn?

According to a 2007 survey of 400 members conducted by the Association of Energy Engineers, energy managers reported an average annual salary of $87,090. Of the respondents, 86 percent reported holding a bachelor's degree, and 37 percent an additional graduate degree. Almost all held more than one professional certification, and more than half had had more than 20 years' experience in the field.

EMCS
Energy Management Control Systems are tools that promotes energy efficiency by accurately monitoring energy consumption and building operations.

That's quite an ambitious program, and it gets even more ambitious with the addition of the Renewable Energy Technician Option, the graduate of which, LCC adds, will:

◆ Appropriately size and recommend renewable energy system types for particular situations.

◆ Understand and put into practice the installation protocol for Photovoltaic and Solar Domestic Hot Water Systems.

How to get there? Well, here's LCC's list of required courses:

First Year
Introduction to Spreadsheets and Databases
Blueprint Reading: Residential and Commercial
Intermediate Algebra
Introduction to Energy Management
Introduction to Sustainability
Residential/Light Commercial Energy Analysis
Alternative Energy Technologies
Introduction to Water Resources
Co-op Ed: Energy Management Seminar
Fundamentals of Physics
English Composition
Air Conditioning Systems Analysis
Energy Efficient Methods
Lighting Fundamentals
Fundamentals of Physics
Human Relations at Work

Second Year
Commercial Air Conditioning Systems Analysis
Lighting Applications
Energy Investment Analysis

MEDIAN SALARIES/WAGES ❖ Energy Management		
Engineer, Facilities Layout	$40,110 –	$92,400
Mechanical Engineer	$52,230 –	$85,320
Environmental Engineer	$55,590 –	$73,660
Electronics Technician	$12.70 – $21.90 / hour	
Certified Electrician	$21.70 – $35.50 / hour	
Civil Engineering Technician	$16.60 – $22.30 / hour	
HVAC Mechanic	$16.00 – $25.00 / hour	
HVAC Field Service Engineer	$23.90 – $30.60 / hour	

SOURCE: WWW.PAYSCALE.COM (DECEMBER 2007)

Technical Writing
Commercial Energy Use Analysis
Energy Control Strategies
Co-op Ed: Energy Management Seminar
Building Energy Simulations
Energy Accounting
Co-op Ed: Energy Management

Add to this program other requirements and a few electives, and you'll see that Lane Community College students are busy—but also very well prepared to enter the workforce at levels commanding initial pay of at least $30,000 a year. ❖

Managing Energy Efficiently: Texas Instruments

In 2004, Texas Instruments embarked on an ambitious project to build the world's first "green," LEED-certified (Leadership in Energy and Environmental Design) semiconductor manufacturing facility in an effort to reduce construction and operating costs and the company's impact on the environment. The new fabrication site (RAFB) was located in Richardson, Texas.

Although building "green" required some additional investment to realize long-term operating benefits, it added up to less than one percent of the construction budget. In addition, the plant was successfully built for an estimated 30 percent less in cost than a similar TI manufacturing plant constructed just 6 miles away only a few years earlier. This latter achievement increased the building's cost competitiveness among other semiconductor manufacturing facilities being built outside of the U.S.

Before any design funding was approved for the construction of the facility, a small group of Texas Instrument employees began investigating sustainable design. They gathered information, compiled data and brainstormed ideas. As the research team began to understand what was possible in their drive toward sustainable design, they knew they needed to solicit management support. A research team member offered TI's senior vice president of Manufacturing a tour of his passive/active solar house. While the tour provided a good primer on sustainable design, it was the low operating cost that really caught the executive's attention. He wanted to know first and foremost, "How much of this design process scales up to a large facility?" The answer: "All of it!" The conversation ended with one last question, "What do you need to make this happen?"

About a month before design funds were approved, more than thirty TIers convened with a dozen folks brought in by Amory Lovins and the Rocky Mountain Institute (RMI). The team held a 3-day design charrette to brainstorm ideas, then analyze and prioritize them. This list was dubbed the "Big Honkin' Ideas." In the end, most of the Big Honkin' Ideas were incorporated along with dozens of other items that came from this meeting.

In addition to sustainable features of the 92-acre site and the utility building, the office building offers four distinct benefits:

Energy Savings

◆ Passive solar orientation minimizes unwanted sunshine and exterior shade screen minimizes summer heat.

◆ Light shelves reduce the need for indoor lighting by bouncing daylight deeper indoors.

◆ Reflective roofing reduces the urban heat island effect.

◆ Quality window glazing provides a balance of good insulation and good visible light transmission.

◆ Smart lighting has built-in motion and photo sensors to respond to indoor conditions.

◆ Solar water heating.

◆ Water turbine-powered hand wash faucets.

Water Savings

◆ Waterless urinals save 40,000 gallons of water each per year.

Improved Air Quality

◆ CO_2 sensor-controlled ventilation provides the intake of fresh air as needed.

◆ Use of safer building materials, including paints, sealants and adhesives.

◆ Locally manufactured materials reduced shipping pollution.

◆ Shuttle buses, free annual mass transit passes, and a carpool matching program will discourage single occupant commuting.

Reduced Material Use

The recycled content of all materials used in building construction is greater than 20 percent:

◆ Ceiling tiles have a recycled material content greater than 80 percent.

◆ The carpet is made from recycled materials and has very low emissions.

◆ Recycling centers for employees.

◆ Use of certified wood and wheatboard ensured the wood was extracted in a sustainable manner and contributes to the preservation of old growth forests.

◆ High-velocity hand dryers conserve paper towels.

SOURCE: TEXAS INSTRUMENTS

Energy Management Resources

ORGANIZATIONS & WEB SITES

American Society of Heating, Refrigeration and Air Conditioning Engineers

ASHRAE, founded in 1894, is an international organization of 50,000 persons. It fulfills its mission of advancing heating, ventilation, air conditioning and refrigeration to serve humanity and promote a sustainable world through research, standards writing, publishing, and continuing education. The organization's web site carries information on all these matters.

www.ASHRAE.org

Department of Energy Federal Energy Management Program

The U.S. government is the largest single energy consumer in the country. The Department of Energy's Federal Energy Management Program (FEMP) works to reduce the cost and environmental impact of the Federal government by advancing energy efficiency and water conservation, promoting the use of distributed and renewable energy, and improving utility management decisions at federal sites.

www.eere.energy.gov/femp

Energy & Environmental Building Association

EEBA provides education and resources to transform the residential design and construction industry to profitably deliver energy efficient and environmentally responsible buildings and communities. Their comprehensive web site offers numerous resources, including the *Builders Guides* that cover four different climates.

www.eeba.org

Energy Management Institute

EMI offers specialized training in management and other aspects of the energy sector. Most of its work is with professionals in petroleum and other fossil fuels, but it has some experience with renewables. Its web site lists courses, workshops, and other professional-development tools.

www.energyinstitution.org

Energy Star Program

Energy Star is a joint program of the U.S. Environmental Protection Agency and the U.S. Department of Energy devoted to energy-efficient products and practices. The program certifies appliances that meet strict energy-efficiency guidelines. Its web site offers information and resources to help plan and undertake projects to reduce energy bills and improve home comfort.

www.energystar.gov

PUBLICATIONS

Clive Beggs, *Energy Management and Conservation* (Butterworth-Heinemann, 2002)
This textbook surveys energy-efficient building practices and issues of cost, supply, taxation and policy. Highly useful for advanced students.

Barney L. Capehart, Wayne C. Turner, and William J. Kennedy, *Guide to Energy Management, 5th edition* (CRC, 2006)
An essential reference for energy managers with chapters written by leading energy professionals on HVAC optimization, improving operations and efficiency ratings of equipment, understanding power factors and energy usage and bills, and the like.

G. G. Rajan, *Optimizing Energy Efficiencies in Industry* (McGraw-Hill Professional, 2002)
A practical textbook full of examples that are useful to students of energy engineering and analysis.

Paul Scheckel, *The Home Energy Diet* (New Society Publishers, 2005)
Learn about measuring, metering, investigating, and considering habits related to household energy use, then how to quantify energy consumption and cost, and how to make informed decisions about cost-effective improvements and upgrades. The book explores the misunderstood concept of efficiency versus cost by comparing fuel costs and equipment choices, including the possibility of using renewable energy for meeting home energy needs.

Albert Thumann and William J. Younger, *Handbook of Energy Audits, 6th edition* (Fairmont Press, 2003)
A comprehensive and practical reference on energy auditing in buildings and industry, containing all the information needed to establish an energy audit program. Accounting procedures, electrical, mechanical, building and process systems analysis, life-cycle costing, and maintenance management are covered in detail.

Donald R. Wulfinghoff, *Energy Efficiency Manual* (Energy Institute, 2000)
If it has to do with energy conservation, this massive (and expensive) handbook has something to say about it. Wulfinghoff, an energy expert, provides scenarios for 400 energy-conservation actions set in homes, offices, commercial buildings, and industrial plants. *Publishers Weekly* says that it "answers just about any question [from] homeowner, plant manager, energy policy guru . . . as practically useful as it is informative."

NOTE: Related resources are listed in the previous chapter (Green Building).

Green Transportation

9   Green Transportation

In 1980, the minimum wage in the United States was $3.10. In 2007, it was $5.85. The buying power of $3.10 in 1980 was, however, worth $7.82 in today's dollars, making today's minimum wage seem—well, not much of a deal.

In 1980, according to the Bureau of Transportation Statistics, the average passenger car got 24.3 miles per gallon. In 2006, that figure had climbed to 30 mpg.

It's comparing apples to oranges, of course, but the second one is a figure we can all be proud of—while at the same time recognizing that there's still much more we can do to improve the fuel efficiency of our vehicles.

The advances of the last quarter-century have three big driving forces, if you'll forgive the pun. The first is the work of smart designers, engineers, and automakers, who figured out ways to make cars leaner and greener without compromising on safety or drivability. (Most of today's cars are in every way improvements on those of 1980 as a result.) The second is the power of government at all levels, municipal to national, to force those efficiencies. The third is the even greater power of the wallet, by which consumers have

Before Tesla, one startup idea Martin and Marc considered was smart sprinkler heads to aid in water conservation. They decided to conserve oil instead. Their Tesla Roadster is a high-performance electric car that can travel about 245 miles on one charge; 0 to 60 mph in less than 4 seconds with a top speed of 125 mph. It uses no gasoline and has zero emissions. *Photo courtesy of Tesla Motors (www.teslamotors.com)*

rewarded companies that have made better vehicles by buying them, even though the cost of a new car in 2006, adding finance charges and taxes, averaged $26,500 for a passenger vehicle or light truck, while, according to the U.S. Department of Energy, in 1980 the figure was $7,600—that is, $19,176 in today's dollars.

The same forces are at work in making green vehicles today, but the opportunities are ever greater. Consumers, automakers, and governments alike are clamoring for better, safer, more fuel-efficient (and, of course, cheaper) vehicles. There's no better time to be involved in the many fields that contribute to that end, and no better time to be conscious of the environmental cost of transportation.

Best known for his work in the television drama series *St. Elsewhere,* Ed Begley Jr. has been an environmental pioneer for decades. Though he lives in the heart of Los Angeles, he keeps his transportation habits simple. "For around L.A.," he tells MSNBC News, "my first mode of transportation is walking; second in the hierarchy is my bicycle; third would be public transportation; and then fourth would be my electric car, for picking up large amounts of groceries, taking my Begley's Best [environmentally friendly housecleaning] products around to be delivered somewhere; and then fifth in the line would be my wife's Prius."

Careers in Green Transportation

Consider how important transportation is in our daily lives. Rightly or wrongly, we rely on long-distance transport to bring us much of what we eat and wear. We live in towns and cities where, for almost all of us, at least a few miles separate our homes from our workplaces, schools, shops, and other important points in our personal geographies. We've grown up in an era in which the automobile takes a place of pride in the very center of existence: it's not just a tool, but a means of personal liberation, self-fulfillment, escape, and fun.

Now, think about a green future in which cars run clean—or at least cleaner. Those of us who still use cars will find this a blessing, of course. But, like Ed Begley Jr., we'll have learned to walk again, having remade our cities so that we can use our feet to get from one place to another. We'll take public transportation. We'll ride our bikes.

And some of us, of course, will take our hydrogen-powered vehicles down to the race track. . . .

In that future, as one energy analyst notes, there'll be plenty of transportation-related jobs:

- ◆ Car and truck mechanic jobs
- ◆ Production jobs
- ◆ Bicycle repair and bike delivery services
- ◆ Gas-station jobs related to biodiesel
- ◆ Public transit jobs related to driving, maintenance, and repair
- ◆ Energy retrofits to increase energy efficiency and conservation
- ◆ Alternative energy equipment installation
- ◆ Manufacturing jobs related to large-scale production of appropriate technologies
- ◆ Materials reuse . . .

The list goes on and on. For the near future, most work in automotive engineering and design takes place within the automotive industry itself, with jobs typically requiring college or technical-school training. For a list of some of those schools, see the following section.

Researchers are needed in many different fields related to transportation. At the National Renewable Energy Laboratory, for instance, the Advanced Vehicle Systems Group conducts R&D on ancillary loads reduction, vehicle systems analysis, energy storage, power and propulsion systems, and advanced power electronics. The Technology Integration & Utilization Group evaluates the performance of new vehicle and fuel technologies as they move into commercial use, and works with industry to develop technical

Top: Sunline's SunBus utilizes the most advanced fuel cell technology available. *Photo: Sunline.org*
Bottom: Hybrid electric trolley uses a 1.9 liter Volkswagen engine and maintenance-free lead acid batteries. *Photo: Leslie Eudy*

solutions and support to overcome problems that may result. And the Fuels Performance Group applies its expertise in fuels, lubricants, and emission control system technologies to guide relevant R&D in support of activities such as the Advanced Petroleum Based Fuels project.

Add to these technical and research fields the need for road and city planners, policy specialists, environmental specialists to conduct environmental-impact studies and to monitor environmental compliance, lawyers, managers, accountants, technical writers and editors, public-relations specialists, advocates, and salespeople—and that list goes on, too—and you'll see that there is much room in green transportation for a wide variety of skills, talents, and wishes.

The Concept Chevy Volt is unlike any other electric vehicle. Its E-Flex system uses a High Energy Battery and range-extending onboard power source that can be configured to run on gas, ethanol or bio-diesel to recharge the battery.
Photo: General Motors

Knowledge Is Power

Automotive engineering is a highly specialized enterprise, and it is no surprise that one of the flagship schools for it is the University of Michigan, whose Ann Arbor campus lies in the heartland of the automotive empire Henry Ford founded a century ago. "We are tasked with the most important objective of the transportation community: to educate the next generation automotive engineers who will achieve the changes required to survive in this competitive, global industry," writes Automotive Engineering Program director Margaret Wooldridge. "We strive to provide the best education on the most current topics important to the future of automotive engineering. Our goal is to create the individuals who will be the instruments of change." For an overview of UM's rigorous program, see *www.interpro-academics.engin.umich.edu/auto/courses.htm*.

Many universities and community colleges offer programs in

automotive engineering and automotive engineering technology. *U.S. News and World Report* gives these schools high marks:

As well, the University of Cincinnati (*www.uc.edu*) offers a new program in automotive design. Students there recently redesigned GM's EV1 (electrical vehicle 1), the first-generation electric car, which aficionados such as Ed Begley Jr. have been eager to see return to the consumer market.   ❖

MEDIAN SALARIES/WAGES ❖ Green Transportation		
Automotive Engineer	$69,390	– $83,400
Transportation Engineer	$42,900	– $62,030
Transportation Planner	$37,450	– $62,480
Traffic Analyst	$43,990	– $68,420
Automotive Mechanic	$14.10 – $21.40 / hour	
SOURCE: WWW.PAYSCALE.COM (DECEMBER 2007)		

Top: A technician performs an inspection of an aftermarket conversion installation. *Photo: Natural Fuels Corporation*
Bottom: NREL engineers test a lithium-ion battery cell to improve performance of electric and hybrid vehicles. *Photo: Warren Gretz, NREL*

Green Transportation Resources

Advanced Vehicles & Fuels Research (NREL)

Working in partnership with public and private organizations, NREL researches, develops and demonstrates innovative vehicle and fuel technologies that reduce the nation's dependence on imported oil and improve our energy security and air quality.

www.nrel.gov/vehiclesandfuels/

Electric Drive Transportation Association

EDTA is dedicated to advancing electric drive as a core technology on the road to sustainable mobility. Their membership includes vehicle and equipment manufacturers, energy providers, component suppliers and end users.

www.electricdrive.org

Governor's Green Government Council (Pennsylvania)

The GGGC publishes a web site that sustains hours of browsing about the fascinating series of initiatives and policy positions to make Pennsylvania a greener place to be. The section on green transportation suggests just how much work there is to be done—and the many workers who will be needed to do it.

www.gggc.state.pa.us/plan0607

Natural Resources Defense Council

The NRDC, a prominent environmental lobbying group, has a very useful web site with in-depth articles on energy and transportation issues.

www.nrdc.org/energy

Rocky Mountain Institute

Founded by visionary Amory Lovins, the Rocky Mountain Institute is one of the nation's leading think tanks devoted to energy efficiency and renewable energy sources. The RMI web site offers a broad-ranging set of documents on transportation, available for download.

www.rmi.org

U.S. Department of Energy Vehicles Technology Program

This DOE program focuses on developing more energy efficient and environmentally friendly highway transportation technologies that will enable America to use less petroleum.

www1.eere.energy.gov/ vehiclesandfuels/

Automakers, such as these three, offer a wealth of information:

Ford

www.ford.com/innovation/ environmentally-friendly

General Motors

www.gm.com/explore

Honda

corporate.honda.com/environment

10

Teaching Energy

10 Teaching Energy

Renewable energy is a fascinating subject. Better put, it is a fascinating set of subjects, calling on knowledge that ranges from the sciences to communications to economics (and home economics) and beyond, on skills as various as climbing a ladder and driving a nail to aligning a wind-turbine propeller and tracking the sun's path.

Students at an NREL science camp learn about solar energy. *Photo: Dave Parson, NREL*

Bringing this set of subjects to students is at once challenging and easy: Challenging because curricula are only now being developed in many aspects of renewable energy, easy because students at every level find the art, science, and industry of renewable energy to be accessible, interesting, and applicable to their daily lives.

Ideally, a teacher bringing lessons and courses to students will share that view. His or her work is getting a little easier, too, for school districts are increasingly going green, and textbooks and other materials are daily arriving on the scene.

Teachers, regrettably, are a little slower to arrive: there just aren't enough of them, particularly in math and the sciences, and the need for teachers is expected to become greater and greater as retirement

**Five Things
Every High-School
Graduate Should
Know About Energy**

What is electricity and
where does it come
from?

What are the sources
of renewable and non-
renewable energy?

What are the environ-
mental and economic
impacts of our energy
use?

What does it mean to
move from a carbon
based to a hydrogen
economy?

What will the energy
portfolio in the future
look like?

SOURCE: NREL OFFICE OF
EDUCATION

and attrition open up positions nationwide. Indeed, it's estimated that by 2010, more than 2.2 million teaching positions will be un-filled, from elementary school to university.

It takes talent and patience to be a teacher, of course; it also takes knowledge, practice, and skill. The pay is often lower than in other sectors, and the stresses of keeping up with many dozens of students a day can take their toll. So, too, can the rigors of keeping up with new knowledge, changing technologies, and evolving curricula. But, as every teacher can tell you, connecting to that one promising student and setting him or her off on a world-changing course can readily outweigh the negatives—and if we are to change the way we make and consume energy and other resources, it will be thanks to bright students pointed in the right direction by excellent teachers.

What follows are some resources, many of which can be adapted to early learners as well as collegians. Whether you're a teacher or a student, you'll find plenty to engage your interest here. ❖

Requirements for Teaching Energy

Middle school teachers (grades 4 – 8) and high school teachers generally specialize in a specific subject, with renewable energy falling under the various sciences, such as earth and space science, life science, physical science, as well as technology / engineering. History, economics and social sciences are also applicable subjects for renewable energy and sustainable living.

Traditionally public school teachers were required to have at least a bachelor's degree, complete an approved teacher education program, and be licensed. Many states, however, are now offering alternative licensing programs to attract people who have bachelor's degrees in a particular subject, but lack the education courses required for a regular license. Private school teachers often have slightly different requirements.

Educators at colleges and universities usually have master's and doctorate degrees, while vocational/technical instructors typically need work-related experience, and often a license or certificate, in their field.

Department of Energy Supports Education

NREL's Education Programs

"By 2050, the world will be struggling to find an additional 20 terawatts of energy and the U.S. will be competing along with countries such as China and India to satisfy our growing energy needs. In order to meet the challenges of this new energy future, we will need new ways of thinking about and producing energy. Here at NREL, we believe that performing renewable energy and energy efficiency R&D is only part of the equation. Education is key to creating a new energy future.

"From elementary school science mentoring to senior-level research participant programs, NREL's educational opportunities help provide the link to this new future. Our K-12 science programs engage young minds in renewable energy and also provide support for teachers. Our college and post-graduate programs help to develop a capable and diverse workforce for the future through mentored research internships and fellowships. And our more senior-level programs—from post-doctoral researchers up to sabbatical and faculty appointments—provide the opportunity to participate in the Laboratory's research and development programs, initiate new areas of research, and establish a base for ongoing collaborations between NREL and our stakeholders."

www.nrel.gov/education

Expanding Your Horizons introduces middle school young women to successful female role models in hands-on workshops. *Photo: Dave Parson, NREL*

Academies Creating Teacher Scientists

The Department of Energy Academies Creating Teacher Scientists (DOE ACTS) program is designed by the Office of Science to create a cadre of outstanding science and math teachers with the proper content knowledge and scientific research experience to serve as leaders and agents of positive change in their local and regional teaching communities. This three-year program uses the unmatched wealth of mentoring talent at the DOE National Laboratories to guide and enrich the teachers' understanding of the scientific and technological world.

www.scied.science.doe.gov/scied/LSTPD/about.htm

MEDIAN SALARIES/WAGES ❖ Teaching Energy		
Teacher, High School	$42,350	– $50,590
Teacher, Middle School	$29,720	– $42,090
Teacher, Elementary School	$32,100	– $53,980
Teacher, College/University	$40,990	– $84,350
Consultant, Education/Training	$49,990	– $75,950
Teacher Assistant		$9.00 – $12.40 / hour

SOURCE: WWW.PAYSCALE.COM (DECEMBER 2007)

Educational Outreach:
Taking Science and Renewable Energy On The Road

A group of students gather around the Generator Bike exhibit to learn about energy creation and use.
Photos: Leonardo on Wheels

Across the state of Utah you can find The Leonardo on Wheels bringing science and documentary arts activities to schools and libraries. The LOW is a traveling science center designed to help students learn, and help teachers teach. "Students love to learn science when they are allowed to experience it. Even familiar principles are given new life when students have the option to learn through inquiry and hands-on discovery," comments Joe Andrade, Director of the Utah Science Center at The Leonardo. *www.theleonardo.org*

Another excellent educational outreach program is RnE2EW —a partnership between the Department of Energy, the National Renewable Energy Laboratory, and BP America. This traveling program brings renewable energy and energy efficiency sciences to students, teachers, and the community. Education staff conducts educational outreach events, hands-on activities, and professional development seminars at schools, conventions, competitions, trade shows, fairs and other exhibitions for students, teachers and consumers.
www.rne2ew.org

Education Resources

Alliance to Save Energy
This organization offers educators a wide range of tools and resources to bring energy efficiency into the classroom while helping students build vital real-world skills. Its web site offers hundreds of lesson plans.
www.ase.org

Discovery Channel Curriculum Center
A trove of information for students and teachers alike, their web site has a curriculum center with lesson plans in all sorts of energy-related fields.
school.discoveryeducation.com/ curriculumcenter

Energy Efficiency and Renewable Energy
This is a superb source of materials for teaching about energy. Students who come to the site can also find ideas for energy-related science projects and resources to help them write reports on energy topics. High-school and college students can also find information on energy careers, energy-related college degrees and programs, scholarships, energy-related internships, and job listings.
U.S. Department of Energy
www1.eere.energy.gov/education

Energy Information Administration
EIA offers a wealth of information on all aspects of renewable energy, including reports and resources for professionals and students alike. The web site contains links to energy-education materials for students and teachers.
www.eia.doe.gov

Energy Quest
This award-winning energy-education web site of the California Energy Commission offers abundant resources for budding scientists, inventors, and energy consumers, who will find the web site full of information on energy production and conservation. Of particular use is the section devoted to student-teacher resources.
www.energyquest.ca.gov

Florida Solar Energy Center
FSEC works with educators to engage and interest students of all levels to help create a brighter energy future. They offer several K-12 curriculum units online, as well as continuing education coursework for teachers.
1679 Clearlake Road
Cocoa, FL 32922-5703
(321) 638-1000
www.fsec.ucf.edu

General Motors K-12 Education Initiatives
GM's educational web site has separate, age-appropriate sections for students and was designed to

provide educators with easy access to GM's educational initiatives and partnerships, and to provide materials for use in the classroom. The web site won the Best Education web site award from the Web Marketing Association in 2004.

www.gm.com/explore/education/

Green Teacher Magazine

A magazine devoted to environmental and global education across the curriculum at all grade levels. Published quarterly, each issue includes some 50 pages of ideas and activities.

P.O. Box 452
Niagara Falls, NY 14304-045
(416) 960-1244
www.greenteacher.com

National Energy Education Development Project (NEED)

NEED offers newsletters, the monthly publication Career Currents, classroom kits, and many other resources. Its web site is a mine of good information.

8408 Kao Circle
Manassas, VA 20110
(703) 257-1117
www.need.org

National Energy Foundation

NEF is a non-profit organization devoted to innovative teacher training and student programs.

www.nef1.org

National Renewable Energy Laboratory Education Program

The NREL Education Program web site is a must-bookmark destination for teachers and students with an interest in all things having to do with energy. The site lists scholarships, grants, internships, and jobs, and it provides abundant resources for lesson planning and curriculum building.

www.nrel.gov/education

National Science Teachers Association

NSTA, with 55,000 members, is a clearinghouse that helps science teachers connect with one another. They provide valuable resources, including lesson plans and reference books.

www.nsta.org

Solar in the Schools

Solar Energy International's SIS program targets grade school youth and focuses on experiential learning of energy concepts. SIS emphasizes critical thinking skills by exploring energy choices, costs and solutions within communities.

PO Box 715
Carbondale, CO 81623
(970) 963-8855
www.solarenergy.org/programs/ solarschools.html

Teachers First

A well-organized site that contains lesson plans and other materials on nearly every topic that touches on renewable energy.

www.teachersfirst.com

Nuclear Electric Power | A Nonrenewable Alternative

11 Nuclear Electric Power | A Nonrenewable Alternative

Two generations ago, nuclear power was heralded as an ideal way to generate electricity—so abundantly, one excited politician said, that it would be "too cheap to meter." Heavily subsidized and heavily promoted, nuclear facilities boomed in the late 1960s and throughout the 1970s. Then, toward the end of the decade, thanks to nuclear scares at the Diablo Canyon plant in California and Three Mile Island in Pennsylvania, coincidentally forecast by the hit movie *The China Syndrome*, construction slowed and came to a near standstill. The economic recession of the time had an effect, too, since electricity from a nuclear plant was and remains twice as expensive as coal- or gas-fired electricity, when all costs are taken into account.

A 2-megawatt photovoltaic system is co-located with the nuclear power operation at Sacramento Municipal Utility District's Rancho Seco Facility. *Photo: Warren Gretz, NREL*

Current concerns about the environment and the future availability of petroleum and other fossil fuels have brought nuclear power back into the picture. In strictest terms, it is an alternative but not a renewable form of energy, and thus, in strictest terms,

Nuclear Facts

A large nuclear plant can generate about a million kilowatts of electricity.

There are about 100 licensed nuclear power reactors in the United States and about 400 in the world.

out of place in this book. For many reasons, nuclear power is also controversial. Still, it is a part of the energy mix that is likely to become more important, at least in the near term, until other, truly renewable forms of energy production are brought online.

Careers in Nuclear Electric Power

According to the federal Bureau of Labor Statistics, even though no new nuclear plants are being built at present, the U.S. nuclear industry will need 90,000 new professional and craft workers in the next decade to replace workers who are retiring or otherwise leaving the workforce.

About 16,000 positions in that workforce are held by nuclear engineers, who operate and support nuclear-energy systems and develop specialized power plants of the sort that are aboard satellites and other spacecraft. They may also work on the nuclear fuel cycle—the production, handling, and use of nuclear fuel and the safe disposal of nuclear waste—or on the development of fusion energy. About half of nuclear engineers are employed by municipal and private utilities, with the rest employed by private firms, universities, and the federal government. Median annual earnings of nuclear engineers were $81,350 in 2007; at the entry level, a nuclear engineer with a bachelor's degree started at an average of $56,000. There is a growing shortage of nuclear engineers—and, for that matter, of all kinds of engineers within the nuclear industry, among them fuel engineers, who specialize in working with the reactor core, and chemical, civil, and structural engineers, who work with various structures inside a nuclear plant. The job outlook therefore seems very promising.

Nuclear plants employ information-technology specialists, health physicists, chemists, radiation technicians, accountants, business managers, and auxiliary staff. Most such positions require a bachelor's degree at minimum. Reactor operators are often but not always college-trained and typically have backgrounds in math, science, or engineering. These operators are licensed by the Nuclear Regulatory Commission after passing a rigorous exam (see *www. nrc.gov/reactors/operating/licensing.html* for details).

Skilled workers employed by nuclear-power plants include electricians, welders, pipefitters, machinists, carpenters, and heavy-equipment operators, trades usually learned on the job, often following training in community colleges or trade schools.

More than other utilities, nuclear plants also require trained security specialists and firefighters, much of whose training comes on the job, and often begins in the military.

A Glance at How Nuclear Power Works

Nuclear power can come from the fission of uranium, plutonium or thorium or the fusion of hydrogen into helium. Most plants use enriched uranium, U-235, most of which is mined in the western United States.

During nuclear fission, a small particle called a neutron hits the uranium atom and it splits, releasing a large amount of energy as heat and radiation. More neutrons are also released. These neutrons go on to bombard other uranium atoms, and the process repeats itself over and over again. This is called a chain reaction.

SOURCE: ENERGY INFORMATION ADMINISTRATION

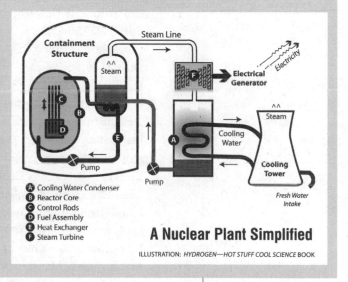

A Cooling Water Condenser
B Reactor Core
C Control Rods
D Fuel Assembly
E Heat Exchanger
F Steam Turbine

A Nuclear Plant Simplified

ILLUSTRATION: *HYDROGEN—HOT STUFF COOL SCIENCE* BOOK

Knowledge Is Power

At present, a small number of community colleges, colleges, and universities offer programs in nuclear engineering—small enough, in fact, that a list follows.

Scholarships are widely available, thanks to efforts by the National Academy for Nuclear Training and the federal government, among other entities. Visit *www.ans.org/honors/internships* for more information. ❖

Air Force Institute of Technology
Arkansas Tech University
Bismarck State College
Bloomsburg University
Central Virginia Community College
Clemson University
Colorado State University
Columbia University
Cornell University
Duke University
Excelsior College
Francis Marion University
Georgia Institute of Technology
Idaho State University
Kansas State University
Lakeland Community College
Lakeshore Technical College
Linn State Technical College
Louisiana State University
Massachusetts Institute of Technology
Massachusetts Maritime Academy
North Carolina State University

Ohio State University
Old Dominion University
Oregon State University
Pennsylvania State University
Purdue University
Rensselaer Polytechnic Institute
Texas A&M University
Three Rivers Community College
U.S. Military Academy
U.S. Naval Academy
University of Arizona
University of California at Berkeley
University of Cincinnati
University of Florida
University of Idaho
University of Illinois at Urbana-Champaign
University of Maryland
University of Massachusetts-Lowell
University of Michigan
University of Missouri at Columbia
University of Missouri at Rolla
University of Nevada Las Vegas
University of New Mexico
University of North Texas
University of Rhode Island
University of South Carolina
University of Tennessee at Knoxville
University of Texas at Austin
University of the District of Columbia
University of Wisconsin at LaCrosse
University of Wisconsin at Madison
Worcester Polytechnic Institute

MEDIAN SALARIES/WAGES ❖ Nuclear Power		
Nuclear Engineer	$55,550	– $97,210
Health Physicist	$34,320	– $79,600
Chemist	$35,180	– $60,000
Research Engineer, Nuclear Fuel	$71,000	– $138,480
Reactor Operator	$24.20 – $41.70 / hour	
Radiation-Protection Engineer	$21.60 – $34.80 / hour	

SOURCE: WWW.PAYSCALE.COM (DECEMBER 2007)

Nuclear Resources

ORGANIZATIONS & WEB SITES

American Nuclear Society

ANS, active on many college campuses, offers good information on careers within the nuclear industry, including career opportunities for those without scientific training.
www.ans.org/pi/edu/students/careers

Nuclear Energy Institute

NEI is the policy organization of the nuclear energy and technologies industry and participates in both the national and global policy-making process. Its objective is to ensure the formation of policies that promote the beneficial uses of nuclear energy and technologies in the United States and around the world. The NEI web site has useful information on all aspects of the industry.
1776 I Street NW, Suite 400
Washington, DC 20006–3708
(202) 739-8000
www.nei.org

U.S. Department of Energy Office of Nuclear Energy

A clearinghouse for information about nuclear power, the ONE web site also includes information about scholarships, internships, and careers.
www.ne.doe.gov

PUBLICATIONS

David Bodansky, *Nuclear Energy: Principles, Practices, and Prospects* (Springer, 2004)
This book presents a view of nuclear energy as an important carbon-free energy option. It discusses the nuclear fuel cycle, the types of reactors used today and proposed for the future, nuclear waste disposal, reactor accidents and reactor safety, nuclear weapon proliferation, and the cost of electric power. It also addresses designs for next-generation reactors, weapons proliferation and terrorism threats, and the potential of alternatives to nuclear energy.

Richard E. Faw, *Fundamentals of Nuclear Science and Engineering* (CRC, 2002)
This introductory textbook explores the fundamentals of nuclear science; the second half of the book introduces the theory of nuclear reactors and its application in electrical power production and propulsion. Highly useful reading for students embarking on the study of nuclear engineering and technology.

Alan M. Herbst and George W. Hopley, *Nuclear Power Now* (Wiley, 2006)
Energy experts Alan Herbst and George Hopley argue that the economic and environmental aspects of nuclear power are largely favorable, that the safety record of power plants is ever improving, and that energy independence is impossible without nuclear energy.

Thank you
to all the companies
and organizations who
so generously supplied
photos and graphics
for this book.
To David Petroy,
Michael Sykes and
Scott Sklar for sharing
their hard-earned
insights. And also to
Linda Lung, Joe Jordan
and Rex Ewing for
their technical advice
regarding the many
facets of renewable
energy.

Appendix

Where to Study: 50 Schools

According to the Digest of Education Statistics, published annually by the U.S. Department of Education, as of the end of the 2006 there were 4,276 degree-granting public and private community colleges, colleges, and universities in the United States. California led the nation in the number of these institutions with 401 of them, followed, in order, by New York, Pennsylvania, Texas, and Ohio. Add to them the nation's 2,297 vocational schools, and that gives a student with an interest in pursuing studies in renewable energy a wealth of choices: 6,573 of them, to be exact, a number that's almost certain to grow along with the nation's population.

What to make of all these choices? Many factors will go into where you decide to study. Do you want to work as a technician? If so, a two-year college, trade school, or apprenticeship will probably fill your needs very well. Do you want to conduct scientific research in, say, hydrogen fuel cell technology? If so, you'll almost certainly need a doctorate in engineering, materials science, or a related field. Would you like to be an evangelist for wind power? You may find that training in sales, business, and communications comes in handy, as well as a solid grounding in how wind turbines work. Such questions are important. So are others, such as how much time you'd like to spend in school, how much education you can afford, where you'd like to study and live, and where the best job opportunities are. The answers to them will have much bearing on your course.

In the preceding chapters, we've provided pointers to some of the best web sites and other resources in individual fields such as geothermal energy and green building. Here we'll list fifty institutions that have distinguished themselves in some aspect or another of renewable energy. For a more directed search, have a look at the U.S. Department of Education's College Opportunities Online Locator (*nces.ed.gov/ipeds/cool*) and at some of the college guidebooks and trade-school listings online. Don't be afraid to ask for help, either—we can almost guarantee you that your local librarian will be delighted to assist you in your quest for information.

Arizona State University
School of Sustainability
PO Box 87511
Tempe, AZ 85287–2511
(480) 727-6963
schoolofsustainability.asu.edu
　　The newly established School of Sustainability
offers courses in sustainable environmental
resources, resource allocation, energy and material use, engineering, and many other subjects.
ASU's College of Technical and Applied Science
also offers courses in renewable and alternate
energy sources, including fuel cells (see
technology.poly.asu.edu).

Ball State University
Center for Energy Research/Education/Service
Architecture Building, Room 018
Muncie, IN 47306
(765) 285-1135
www.bsu.edu
　　Ball State University offers a special
Interdepartmental Education Program of
Clustered Academic Minors in Environmentally
Sustainable Practices. The minors offered
include Environmental Policy, Sustainable
Land Systems, and Environmental Context for
Business.

Bradley University
1501 W. Bradley Avenue
Peoria, IL 61625
(309) 677-2745
www.bradley.edu
　　An independent, privately endowed university,
Bradley offers undergraduate and graduate programs in a variety of fields. It has a particularly
strong engineering program with outreach to
high schools and intermediate schools throughout the region, and conducts vigorous research
in alternative and renewable energy.

California Institute of Technology
1200 East California Blvd.
Pasadena, CA 91125
(626) 395-6811
www.caltech.edu
　　Caltech is a leading research institution in
the study of renewable energy, especially solar
energy and biofuels. Its chemistry department,
to name just one program, sponsors research in
solar energy conversion and storage, methane
oxidation, and nitrogen fixation.

California Polytechnic State University
College of Agriculture, Food and
　　Environmental Sciences
San Luis Obispo, CA 93407
(805) 756-2161
www.calpoly.edu
　　The BioResource and Agricultural Engineering
major at Cal Poly offers hands-on experience in
a wide range of engineering skill areas, renewable energy and waste treatment, electronics
and control systems, and resource information
systems. The BRAE program is accredited by
the Accreditation Board for Engineering and
Technology (ABET).

Coconino Community College
Lone Tree Campus
2800 S. Lone Tree Road
Flagstaff, AZ 86001–2701
(928) 527-1222, (800) 350-7122
www.coconino.edu
　　CCC offers a certification program as an alternative energy technician. With the abundance
of wind and sun on the Colorado Plateau, it's
well placed for work in the field.

Colorado State University
Warner College of Natural Resources
103 Natural Resources Building
Fort Collins, CO 80523
(970) 491-5629
www.warnercnr.colostate.edu
> Warner College of Natural Resources is committed to offering a comprehensive range of undergraduate and graduate degree programs that directly address today's most important environmental and natural resource issues. Its programs are grounded in state-of-the-art science and technologies and involve students in direct problem-solving experiences. Its students are well prepared to become leaders in environmental and natural resources management and science.

Columbia University
2960 Broadway
New York, NY 10027-6902
(212) 854-1754
www.cheme.columbia.edu
> A leading university with undergraduate and graduate instruction in every field that touches on renewable energy, Columbia is home to a highly ranked Department of Mechanical Engineering that offers courses in energy sources, materials science, electrochemistry, and energy systems.

Cornell University
410 Thurston Avenue
Ithaca, NY 14850-2488
(607) 255-5241
www.cornell.edu
> Cornell offers a range of courses in many areas of interest to students and practitioners of renewable energy. One of particular interest is the Engineering Management program, geared toward engineers who want to stay in a technological environment but also to focus on managerial roles, which are of great importance in project management in developing technologies. Through an in-depth, real-world group design project, and course content in management science, project management, decision and risk analysis, information technology, finance and accounting, systems analysis, and organizational behavior, students gain the technical and managerial skills necessary to become effective engineering managers.

Drexel University
Dept. of Chemical and Biological Engineering
3141 Chestnut Street
Philadelphia, PA 19104
(215) 895-2227
www.chemeng.drexel.edu
> Development of technologies for harnessing energy from renewable resources and for reducing the environmental impact of worldwide resource consumption is now receiving major national and international attention. In the Chemical and Biological Engineering Department at Drexel University, there are several faculty leading innovative research programs related to Energy and the Environment. Examples of research projects include production of biodiesel from natural and waste oils containing high free-fatty-acid contents, developing new materials for higher efficiency solar cells, deriving polymers from renewable fatty acid monomer, and modifying the composition and structure of fuel cell catalyst layers for better performance.

Fitchburg State College
160 Pearl Street
Fitchburg MA 01420–2697
(978) 345-2151
www.fsc.edu
   Fitchburg State College offers several degrees and certificates in energy-related fields, including programs in industrial technology, architectural technology, and facilities management. The architectural technology concentration offers the study of technical systems in architecture, including energy systems, and ends with the development of professional practices.

Florida State University
Energy & Environmental Alliance
361 Bellamy Building
Tallahassee, FL 32306-4016
(904) 644-0707
eea.freac.fsu.edu
   The alliance serves as an energy-education clearinghouse with a comprehensive energy education program for Florida teachers and students. Through the program, a variety of curricula has been developed, including printed materials, CD-ROMs, and videodiscs. Of particular interest to those planning careers involving educating consumers in energy issues.

George Washington University
Institute for the Environment
Rice Hall, Suite 603
2121 Eye Street, NW
Washington, DC 20052
(202) 994-3666
www.gwu.edu
   George Washington University offers several programs of study, both graduate and undergraduate, that focus on the environment and sustainability from different perspectives, including natural science, social science, and health science. The online Environmental Curriculum Guide describes these and other available courses in detail.

Illinois Institute of Technology
Energy and Sustainability Institute
3300 South Federal Street
Chicago, IL 60616-3793
(312) 567-3000
www.iit.edu
   The Energy and Sustainability Institute develops close collaboration among numerous programs at IIT with a focus on development of energy-related interprofessional educational and research programs and proposals. IIT hopes to take a leading position in high-tech renewables as photovoltaic, solar-thermal and wind energy, and the use of hydrogen as the dominant transportation fuel in fuel-cell-powered electric vehicles.

Iowa Lakes Community College
19 South 7th Street
Estherville, IA 51334
(712) 362-2604
www.iowalakes.edu/programs_study/industrial/ wind_energy_turbine/index.htm
   The Wind Energy and Turbine Technology Program is the first in the state of Iowa. Since the number of wind turbines in Iowa is expected to grow significantly, Iowa Lakes Community College is working to help meet the growing demand for skilled technicians who can install, maintain, and service modern wind turbines. This is a two-year program offering an associate's degree in applied science. (A one-year diploma is also available.)

Massachusetts Institute of Technology (MIT)
77 Massachusetts Avenue
Cambridge, MA 02139-4307
(617) 253-1000
web.mit.edu
   MIT offers top-flight programs at the undergraduate and graduate level in nearly every career area touching on renewable energy. It also allows build-it-yourself programs in engineering, sustainability, and other subjects, many of which are available free online.

Michigan State University

College of Agriculture and Natural Resources
102 Agriculture Hall
East Lansing, MI 48824
(517) 355-0232
canr.msu.edu

Michigan State, an old and well-regarded land-grant college, is the home of the Biomass Conversion Research Laboratory. Its agriculture college offers several graduate, undergraduate, and two-year certificate programs in fields that pertain to renewable energy, including biosystems and agricultural engineering, construction management, crop and soil sciences, forestry, and agricultural technology.

Montana Tech of the University of Montana

1300 West Park Street
Butte, MT 59701
(800) 445-8324
www.mtech.edu

Offering two- and four-year degrees in both trade and academic fields, Montana Tech has a well-regarded program in geological engineering and is located in one of America's most active areas for geothermal energy.

North Carolina State University

North Carolina Solar Center
Box 7401
North Carolina State University
Raleigh, NC 27695-7401
(919) 515-5666
www.ncsc.edu

Created in 1988, the North Carolina Solar Center serves as a clearinghouse for solar and other renewable energy programs, information, research, technical assistance, and training for the citizens of North Carolina and beyond. Students at NCSU have opportunities to learn at the center, and the school also offers excellent programs in many other renewable energy technologies.

Oklahoma City Community College

7777 South May Avenue
Oklahoma City, OK 73159-4444
(405) 682-1611
www.okc.cc.ok.us

OKCC's program prepares individuals to apply basic engineering principles and technical skills in support of engineers and other professionals engaged in developing solar-powered energy systems. Includes instruction in solar energy principles, energy storage and transfer technologies, testing and inspection procedures, system maintenance procedures, and report preparation.

Old Dominion University

108 Rollins Hall
Norfolk, VA 23529-0050
(757) 683-3685
www.odu.edu

Old Dominion offers a range of programs in energy-related fields, including one of the nation's few degrees in nuclear engineering technology.

Oregon State University

Corvallis, OR 97331-4501
(541) 737-1000
oregonstate.edu

Oregon State's Energy Resources Research Laboratory (ERRL) has managed the data collection, quality assurance and analysis for the Bonneville Power Administration's wind-energy resource studies since 1978. It maintains a large database of wind data for the Pacific Northwest. OSU offers courses in a broad range of fields relating to renewable energy; its Department of Chemical, Biological, and Environmental Engineering is a national leader in research.

Pennsylvania College of Technology
One College Avenue
Williamsport, PA 17701-5799
(800) 367-9222, (570) 327-4761
www.pct.edu
    PCT offers two- and four-year courses in solar technology, electric-power-generation technology, electrical technology, and other areas relevant to renewable energy.

Rensselaer Polytechnic Institute
Department of Earth and Environmental Sciences
Jonsson-Rowland Science Center, 1W19
110 8th Street
Troy, NY 12180
(518) 276-6474
www.rpi.edu/dept/ees/
    RPI offers undergraduate and graduate training in the geosciences and has become a leading center of geothermal-energy studies. The programs include the study of Earth's component materials, the development of its structures and surface features, the processes by which these change with time, and the origin, discovery, and protection of its resources—water, fuels, and minerals.

San Juan College
4601 College Blvd.
Farmington, NM 87402
(505) 566-3003
www.sjc.cc.nm.us
    The Renewable Energy Program gives the student a solid foundation in the fundamental design/ installation techniques required to work with renewable technologies. The concentration in Photovoltaic System Design and Installation is offered as an A.A.S. degree and/or a one-year certificate. Both avenues include hands-on electrical training in their modern computer-based laboratory setting where the National Electrical Code (NEC) is emphasized throughout curriculum.

Sonoma State University
Environmental Studies and Planning
Rohnert Park, CA 94928
(707) 664-2306
www.sonoma.edu/ensp
    Sonoma State students pursue a concentrated course of study in one of four professionally oriented study plans for the bachelor's degrees: Environmental Conservation and Restoration; Environmental Education; Environmental Technology (with tracks in Energy Management and Design, Hazardous Materials Management, and Water Quality); and Planning. In addition to its academic focus, the department also offers the EarthLab, a one-acre education, demonstration, and research center.

Stanford University
630 Serra Street, Suite 120
Stanford, CA 94305–6032
(650) 723-4291
www.stanford.edu
    One of the finest private universities in the world, Stanford is home to the Global Climate and Energy Project, whose goal is to develop environmentally sustainable fuel technologies that are cost-effective on a large scale. About 225 graduate students, postdoctoral students, and other researchers work on projects ranging from better ways to store hydrogen to radical new ways to make engines more efficient. GCEP's lead scientists assume that no single technology will replace fossil fuels.

Stark State College of Technology
6200 Frank Avenue NW
North Canton, OH 44720
(330) 494-6170
www.starkstate.edu/fuelcell
Stark State College of Technology is the site of Ohio's $4.7 million Fuel Cell Prototyping Center, completed in the fall of 2006. The Center is designed for use by emerging and fuel cell-related technology companies to assist in the prototyping and demonstration stages of the development of fuel cell-based power generation systems. Stark State now offers a fuel cell technology track and scholarship as part of the mechanical engineering technology program.

Syracuse University
Syracuse, NY 13244
(315) 443-1870
www.syr.edu
The Building Energy and Environmental Systems Laboratory (BEESL) in the Department of Mechanical and Aerospace Engineering is a key research lab. Its mission, in part, is to advance the science and develop innovative technologies in the areas of indoor environmental quality and building energy efficiency. BEESL's labs are used for studies of combined air, heat, moisture and contaminant transport through building envelopes; interactions between indoor, outdoor environments and HVAC systems/components; sensitivity, accuracy and reliability of environmental sensors and control systems; and many other topics of interest to energy managers and students of renewable energy.

Texas State Technical College West Texas
650 East Highway 80
Abilene, TX 79601
(325) 672-7091
www.westtexas.tstc.edu
The Wind Energy Technology (WET) Associates Degree Program is designed to provide graduating students with the skills necessary to facilitate an easy transition into many levels of the Wind Energy Industry. Curriculum was developed through the collaborative efforts of subject matter experts from Texas State Technical College West Texas and an advisory board of Wind Energy Managers.

Texas Tech University
Wind Science and Engineering Research Center
10th and Akron
Lubbock, TX 79409
(806) 742-3476
www.wind.ttu.edu
The Wind Science and Engineering Research Center at Texas Tech University is an internationally recognized leader in damage documentation in windstorms, boundary-layer wind, and shelter design. Its faculty and students conduct research and engineering in structures, bridges, wind tunnels, full scale data collection, economics, atmospheric science, mathematics, and many other disciplines.

Thomas Edison State College
101 West State Street
Trenton, NJ 08608-1176
(888) 442-8372
www.tesc.edu
Thomas Edison State College offers associate's and bachelor's degrees in applied science, business administration, and technology.

University of Alaska Southeast
11120 Glacier Hwy.
Juneau, AK 99801
(907) 465-8774
uas.alaska.edu
> The UAS offers a program that teaches the essentials of building diagnostic assessment and building durability, performance, and energy efficiency. The one-year program leads to a certificate in residential building science.

University of California at Los Angeles
Department of Architecture
Los Angeles, CA 90095
(310) 454-7348
www.ucla.edu
> The internationally recognized UCLA architecture program offers a specialization in energy design and climate-responsive design. Credit is given for AIA continuing education; see *www.aud.ucla.edu/energy-design-tools* for more information.

University of California, Merced
PO Box 2039
Merced, CA 95344
(209) 228-4400
naturalsciences.ucmerced.edu
> Merced's undergraduate major in Earth Systems Science is designed to provide students with a quantitative understanding of the physical, chemical, and biological principles that control the processes, reactions, and evolution of Earth. Core courses cover the fundamentals of chemistry, biology, hydrology, ecology, and earth sciences. Graduates are well prepared for either graduate studies or jobs in the areas of environmental science and conservation, ecosystem and natural resource management and science, and many aspects of agricultural sciences.

University of Central Florida
Florida Solar Energy Center
1679 Clearlake Road
Cocoa, FL 32922-5703
(321) 638-1000
www.fsec.ucf.edu
> As the state of Florida's energy research institute, the Florida Solar Energy Center conducts research in building science, photovoltaics, solar thermal, hydrogen and alternative fuels, fuel cells, and other advanced energy technologies. The school offers courses and workshops that take advantage of this renowned resource.

University of Colorado Environmental Center
University Memorial Center, Room 355
Boulder, CO 80309-0024
(303) 492-8308
www.ecenter.colorado.edu
> Colorado is one of the greenest campuses in the nation, literally and figuratively, and a leader in the use of renewable energy sources such as wind and solar power. The university offers a wide range of programs in every aspect of renewable energy.

University of Florida
Florida Energy Extension Service
P.O. Box 110940
Gainesville, FL 32611-0940
(352) 392-5684
www.energy.ufl.edu
> The University of Florida provides continuing education courses in green building for Florida licensed building contractors and architects. "Build Green and Profit" is a two-day program that offers topical modules integrated to provide a whole-house perspective on green construction.

University of Illinois at Chicago

Energy Resources Center
1309 South Halsted Street, 2nd Floor
Chicago, IL 60607
(312) 996-4490
www.erc.uic.edu

> The Energy Resources Center is an interdisciplinary program whose areas include energy management assessments, economic modeling, analysis of policy and regulatory initiatives, and education. ERC provides energy auditor/energy efficiency training to a selected team of high-school students each year, as well as an energy camp that offers a week of more intense training to the same students.

University of Massachusetts

Center for Energy Efficiency and Renewable Energy
160 Governors Drive
Amherst, MA 01003
(413) 545-4359
www.ceere.org

> The Center for Energy Efficiency and Renewable Energy (CEERE) provides technological and economic solutions to environmental problems resulting from energy production, industrial, manufacturing, and commercial activities, and land-use practices. CEERE offers research, training and educational experiences for graduate and undergraduate engineers and scientists.

University of Michigan

Erb Institute for Global Sustainable Enterprise
Dana Building
440 Church Street
Ann Arbor, MI 48109–1041
(734) 647-9799
www.erb.umich.edu

> Students in the Erb Institute MBA/MS program earn both a master of business administration (MBA) from the Ross School of Business and a master of science (MS) from the School of Natural Resources and Environment. The program equips leaders, executives, and managers with the skills and knowledge necessary to create environmentally and economically sustainable organizations.

University of Nevada, Reno

1664 N. Virginia Street
Reno, NV 89557–0042
(775) 784-1110
www.unr.edu

> A solid four-year institution with graduate programs in a wide range of fields, Nevada recently inaugurated a program of study in renewable energy. The program includes students from Truckee Meadows Community College, using the vast, nearby Steamboat Geothermal Complex and Nevada's abundant sunshine as laboratories.

University of New South Wales

School of Photovoltaic and Renewable Energy
     Engineering
Electrical Engineering Building
Sydney NSW 2052
Australia
+61 2 9385 4018
www.pv.unsw.edu.au

The School of Photovoltaic and Renewable
Energy Engineering delivers the world's first
photovoltaic and renewable energy engineering degree programs. The School includes
the Australian Research Council Centre for
Advanced Silicon Photovoltaics and Photonics
(also known as the ARC Photovoltaics Centre of
Excellence).

University of Texas at Austin

Nuclear Engineering Teaching Lab
Pickle Research Campus, R-9000
Austin, TX 78712
(512) 232-2467
www.me.utexas.edu/%7Enuclear/

The Nuclear and Radiation Engineering Program
sees its mission as educating the next generation
of leaders in nuclear science and engineering,
conducting leading research at the forefront of
the national and international nuclear community, and applying nuclear technology for solving
multidisciplinary problems. Its teaching lab
houses the 1.1 MW TRIGA research reactor and
a wide array of neutron beam ports, counting
rooms, and radiochemistry and health-physics
facilities.

University of Wisconsin

Solar Energy Program
1500 Engineering Drive
Madison, WI 53706-1687
(608) 263-1589
sel.me.wisc.edu

For study at the Solar Energy Laboratory,
applicants must have a bachelor's degree in
engineering or its physical science equivalent,
have a strong undergraduate record and be
admitted to either the mechanical or chemical
engineering department. Three letters of recom-
mendation are required. The GRE examination
is recommended.

University of Wyoming

School of Environment and Natural Resources
P.O. Box 3963
Laramie, WY 82701-3963
(307) 766-5080
www.uwyo.edu

The University of Wyoming School of
Environment and Natural Resources provides
degree programs in research and policy aspects
of environmental and natural resource decision-making. University of Wyoming focuses
on six operational clusters: energy; ecosystems;
allocation systems; environmental conditions
and economic and social interactions; land and
water resources; and atmospheric resources and
systems.

Washington University
Department of Energy, Environmental &
    Chemical Engineering
One Brookings Drive, Box 1180
St. Louis, MO 63130
(314) 935-5548
eec.wustl.edu
> This newly created department has a focus on environmental engineering science, energy systems, and chemical engineering. The department provides integrated and multidisciplinary programs of scientific education, as well as participation in cutting-edge research with faculty and industrial partners and access to state-of-the-art facilities and instrumentation. The basic degree is an undergraduate degree in chemical engineering. Graduate degrees (Master of Science and Doctor of Science) are offered in both chemical engineering and environmental engineering science on completion of a course of study and research work. A joint degree program with the School of Law allows interested students to obtain both a J.D. and M.S. in environmental engineering science.

Wayne State University
College of Engineering: Alternative Energy
    Technology
Wayne State University
Detroit, MI 48202
(313) 577-2424 or (877) WSU-INFO
www.wayne.edu
> The Wayne State University master's of science program in alternative energy technology is open to students with a bachelor's degree in engineering, and in other mathematics-based sciences in exceptional cases. Qualified admission is possible if the applicant has significant professional experience.

Washington State University
Energy Program at Olympia
925 Plum St. SE, Building 4
Olympia, WA 98504-3165
(360) 956-2040
www.energy.wsu.edu
> The Energy Program teaches energy and ventilation codes for construction staff and designers.

Yavapai College
Construction Technology Department
2275 Old Home Manor Way
Chino Valley, AZ 86323
(928) 717-7726
yc.edu
> Students learn through hands-on design and construction of a high-performance house using renewable energy. The college offers a one-year advanced certificate or a two-year AAS degree in Architectural Graphics, Residential Building Technology, and Residential Construction Management.

General Reference Web Sites

Additional details about these web sites can be found at the end of various chapters.

Career College Association
www.career.org

Center for Renewable Energy and Sustainable Technology
www.crest.org

Energy Management Institute
www.energyinstitution.org

Energy Quest
www.energyquest.ca.gov

FindSolar.com
www.findsolar.com

Florida Solar Energy Center
www.fsec.ucf.edu

Geothermal Education Office
geothermal.marin.org

National Energy Education Development Project (NEED)
www.need.org

National Energy Foundation
www.nef1.org

National Fuel Cell Research Center
www.nfcrc.uci.edu

Natural Resources Defense Council
www.nrdc.org/energy

North American Board of Certified Energy Practitioners
www.nabcep.org

Ocean Energy, California Energy Commission
www.energy.ca.gov/development/oceanenergy

Renewable Energy Access
www.renewableenergyaccess.com

Rocky Mountain Institute
www.rmi.org

SustainableBusiness.com
sustainablebusiness.com

Online Job Listings / Workshops & Hands-On Training

Additional details about these web sites can be found at the end of various chapters.

Environmental Career Center
www.environmentalcareer.com

Environmental Career Opportunities
www.ecojobs.com

Environmental Careers Organization
www.eco.org

Green Energy Jobs
www.greenenergyjobs.com

Greenjobs
www.greenjobs.com

Midwest Renewable Energy Association
www.the-mrea.org

Renewable Energy Access
www.renewableenergyaccess.com

Solar Energy International
www.solarenergy.org

Solar Living Institute
www.solarliving.org

SustainableBusiness.com
sustainablebusiness.com

Yestermorrow Design/Build School
www.yestermorrow.org

NOTE: Many professional societies and associations post jobs, as do major employers within each field. Check their web sites for details.

Government Reference Web Sites

Additional details about these web sites can be found at the end of various chapters.

Energy Information Administration
www.eia.doe.gov

Energy Star Program
www.energystar.gov

National Renewable Energy Laboratory
Advanced Vehicles & Fuels, Biomass,
Buildings, Geothermal, Hydrogen & Fuel Cells,
Solar, Wind
www.nrel.gov

National Renewable Energy Laboratory:
Education
www.nrel.gov/education

U.S. Department of Energy
Energy Efficiency and Renewable Energy
Biomass, Building Technologies, Federal Energy
Management, Geothermal, Hydrogen & Fuel
Cells, Solar Energy, Vehicle Technologies, Wind
& Hydropower
eere.energy.gov

U.S. Environmental Protection Agency
www.epa.gov/opptintr/greenbuilding

Professional Associations & Societies

Additional details about these web sites can be found at the end of various chapters.

American Institute of Architects
www.aiaonline.com

American Bioenergy Association
www.biomass.org

American Hydrogen Association
www.clean-air.org

American Society of Heating, Refrigeration and Air Conditioning Engineers
www.ASHRAE.org

American Solar Energy Society
www.ases.org

American Wind Energy Association
www.awea.org

Association of Energy Engineers
www.aeecenter.org

Electric Drive Transportation Association
www.electricdrive.org

Energy and Environmental Building Association
www.eeba.org

European Wind Energy Association
www.ewea.org

Geothermal Energy Association
www.geo-energy.org

Geothermal Resources Council
www.geothermal.org

International Association for Hydrogen Energy
www.iahe.org

International Solar Energy Society
www.ises.org

National Association of Home Builders
www.nahb.org

National Hydrogen Association
www.hydrogenassociation.org

National Hydropower Association
www.hydro.org

National Science Teachers Association
www.nsta.org

National Society of Professional Engineers
www.nspe.org

Solar Energy Industries Association
www.seia.org

Sustainable Buildings Industry Council
www.SBICouncil.org

U.S. Fuel Cell Council
www.usfcc.com

U.S. Green Building Council
www.usgbc.org

State Energy Offices

Alabama Energy Office
Dept. of Economic and
    Community Affairs
401 Adams Avenue
P.O. Box 5690
Montgomery, AL 36103-5690
Phone: (334) 242-5292
Fax: (334) 242-0552
www.adeca.alabama.gov

Alaska Energy Authority
Alaska Industrial Development
    and Export Authority
813 W. Northern Lights Blvd.
Anchorage, AK 99503
Phone: (907) 771-3000
Fax: (907) 771-3044
www.aidea.org/aea/index.html

Arizona Energy Office
Arizona Dept. of Commerce
1700 W. Washington, Suite 600
Phoenix, AZ 85007
Phone: (602) 771-1194
Fax: (602) 771-1203
www.azcommerce.com/Energy/

Arkansas Energy Office
Arkansas Economic Development
    Commission
One Capitol Mall
Little Rock, AR 72201
Phone: (501) 682-6103
Fax: (501) 682-7394
www.1800arkansas.com/Energy/

**California Energy
    Commission**
Renewable Energy Office
1516 Ninth Street, MS-29
Sacramento, CA 95814-5512
Phone: (916) 654-4287
Fax: (916) 654-4420
www.energy.ca.gov

Colorado Energy Office
Governor's Energy Office
225 East 16th Avenue, Suite 650
Denver, CO 80203
Phone: (303) 866-2100
Fax: (303) 866-2930
www.colorado.gov/energy/

Connecticut Energy Office
Connecticut Office of Policy and
    Management
450 Capitol Ave
Hartford, CT 06106-1379
Phone: (860) 418-6374
Fax: (860) 418-6496
www.ct.gov/opm

Delaware Energy Office
Department of Natural
    Resources and Environmental
    Control
1203 College Park Drive, Ste. 101
Dover, DE 19904
Phone: (302) 735-3480
Fax: (302) 739-1840
www.delaware-energy.com

**District of Columbia Energy
    Office**
District Department of the
    Environment
2000 14th Street NW, Suite 300E
Washington, DC 20009
Phone: (202) 673-6718
Fax: (202) 673-6725
www.dceo.dc.gov

Florida Energy Office
Florida Dept. of Environmental
    Protection
2600 Blair Stone Road MS-19
Tallahassee, FL 32399
Phone: (850) 245-8002
Fax: (850) 245-8003
www.dep.state.fl.us/energy/

**Georgia Division of Energy
    Resources**
Georgia Environmental Facilities
    Authority
233 Peachtree Street NE
Harris Tower, Suite 900
Atlanta, GA 30303
Phone: (404) 584-1000
Fax: (404) 584-1069
www.gefa.org/

Hawaii Energy Office
Energy Resources and Technology
   Division
235 S. Beretania St., Room 502
P.O. Box 2359
Honolulu, Hawaii 96804-2359
Phone: (808) 587-3807
Fax: (808) 586-2536
www.hawaii.gov/dbedt/
   info/energy

Idaho Energy Division
Idaho Dept. of Energy Resources
322 E. Front Street
P.O. Box 83720
Boise, Idaho 83720-0098
Phone: (208) 287-4800
Fax: (208) 287-6700
www.idwr.idaho.gov/energy/

Illinois Energy Bureau
Illinois Dept. of Commerce and
   Economic Opportunity
620 East Adams Street
Springfield, IL 62701
Phone: (217) 782-7500
Fax: (217) 785-2618
www.ilbiz.biz/dceo/Bureaus/
   Energy_Recycling/

Indiana Energy Office
Department of Commerce
101 W. Ohio Street, Suite 1250
Indianapolis, IN 46204
Phone: (317) 232-8939
Fax: (317) 232-8995
www.in.gov/energy/

Iowa Energy Bureau
Iowa Dept. of Natural Resources
Wallace State Office Building
502 East 9th Street
Des Moines, IA 50319
Phone: (515) 281-5918
Fax: (515) 242-5818
www.iowadnr.gov/energy/

Kansas Energy Programs
Kansas Corporation Commission
1500 SW Arrowhead Road
Topeka, Kansas 66604-4027
Phone: (785) 271-3170
Fax: (785) 271-3268
www.kcc.state.ks.us/energy/
   index.htm

Kentucky Division of Energy
500 Mero Street, 12th Floor
Frankfort, KY 40601
Phone: (502) 564-7192
Fax: (502) 564-7484
www.energy.ky.gov

Louisiana Energy Office
Technology Assessment Division
   Dept. of Natural Resources
P.O. Box 44156
617 North Third Street
Baton Rouge, LA 70802
Phone: (225) 342-1399
Fax: (225) 342-1397
www.dnr.state.la.us

Maine Division of Energy
   Programs
Maine Public Utilities Commission
State House Station No. 18
Augusta, ME 04333-0018
Phone: (207) 287-3318
Fax: (207) 287-1039
www.efficiencymaine.com

Maryland Energy Office
Maryland Energy Administration
1623 Forest Drive, Suite 300
Annapolis, MD 21403
Phone: (410) 260-7511
Fax: (410) 974-2250
www.energy.state.md.us

Massachusetts Division of
   Energy Resources
Dept. of Economic Development
100 Cambridge Street, Suite 1020
   Boston, MA 02114
Phone: (617) 727-4732
Fax: (617) 727-0030
www.magnet.state.ma.us/doer

Michigan Energy Office
Michigan Public Service
   Commission
P.O. Box 30221
Lansing, MI 48909
Phone: (517) 241-6030
Fax: (517) 241-6101
www.michigan.gov/mpsc

Minnesota Energy Division
Minnesota Dept. of Commerce
85 7th Place East, Suite 500
St. Paul, MN 55101-2198
Phone: (651) 297-2545
Fax: (651) 297-7981
www.commerce.state.mn.us

Mississippi Energy Division
MS Development Authority
P.O. Box 849
510 N. West Street, 6th Floor
Jackson, MS 39205
Phone: (601) 359-6600
Fax: (601) 359-2314
www.mississippi.org
   Look under Programs / Energy

Missouri Energy Center
Dept. of Natural Resources
P.O. Box 176
Jefferson City, MO 65102
Phone: (573) 751-2254
Fax: (573) 751-6860
www.dnr.mo.gov/energy/

Montana Energy Office
Dept. of Environmental Quality
P.O. Box 200901
1100 North Last Chance Gulch
    Room 401-H
Helena, MT 59620-0901
Phone: (406) 841-5240
Fax: (406) 841-5091
www.deq.mt.gov/energy/

Nebraska Energy Office
P.O. Box 95085
1111 "O" Street, Suite 223
Lincoln, NE 68509-5085
Phone: (402) 471-2867
Fax: (402) 471-3064
www.neo.ne.gov

Nevada Energy Office
Dept. of Business and Industry
727 Fairview Drive, Suite F
Carson City, NV 89701
Phone: (775) 687-9700
Fax: (775) 687-9714
http://energy.state.nv.us

**New Hampshire Office of
    Energy and Planning**
4 Chenell Dive
Concord, NH 03301
Phone: (603) 271-2155
Fax: (603) 271-2615
www.nh.gov/oep

**New Jersey Office of Clean
    Energy**
NJ Board of Public Utilities
P.O. Box 350
44 S. Clinton Avenue
Trenton, NJ 08625-0350
Phone: (609) 777-3335
Fax: (609) 777-3330
www.bpu.state.nj.us

New Mexico Energy Office
New Mexico Energy, Minerals
    and Natural Resources Dept.
1220 S. St. Francis Drive
P.O. Box 6429
Santa Fe, NM 87505
Phone: (505) 476-3310
Fax: (505) 476-3322
www.emnrd.state.nm.us/ecmd

**New York State Energy
    Research and Development
    Authority**
17 Columbia Circle
Albany, NY 12203
Phone: (866) NYSERDA
Fax: (518) 862-1091
www.nyserda.org

North Carolina Energy Office
NC Dept. of Administration
1830A Trillery Place
Raleigh, NC 27604
Phone: (919) 733-2230
Fax: (919) 733-2953
www.energync.net

North Dakota Energy Office
Division of Community Services
North Dakota Dept. of Commerce
P.O. Box 2057
1600 E. Century Avenue, Suite 2
Bismarck, ND 58502-2057
Phone: (701) 328-5300
Fax: (701) 328-2308
www.state.nd.us/dcs/Energy

**Ohio Office of Energy
    Efficiency**
Ohio Dept. of Development
77 South High Street, 26th Floor
Columbus, OH 43216
Phone: (614) 466-6797
Fax: (614) 466-1864
www.odod.state.oh.us/cdd/oee/

Oklahoma Energy Office
Oklahoma Dept. of Commerce
P.O. Box 26980
6601 North Broadway
Oklahoma City, OK 73126
Phone: (405) 815-5347
Fax: (405) 815-5344
www.okcommerce.gov

Oregon Energy Office
625 Marion Street, N.E.
Salem, OR 97301
Phone: (503) 378-4040
Fax: (503) 373-7806
www.oregon.gov/energy

Pennsylvania Energy Office
Department of Environmental
    Protection
P.O. Box 2063
400 Market Street, RCSOB
Harrisburg, PA 17105-2063
Phone: (717) 783-0542
Fax: (717) 783-2703
www.dep.state.pa.us

Rhode Island Energy Office
One Capital Hill, 2nd Floor
Providence, RI 02908
Phone: (401) 574-9100
Fax: (401) 574-9125
www.riseo.state.ri.us

South Carolina Energy Office
1201 Main Street, Suite 430
Columbia, SC 29201
Phone: (803) 737-8030
Fax: (803) 737-9846
www.energy.sc.gov

South Dakota Energy Office
Governor's Office of Economic
    Development
711 E. Wells Avenue
Pierre, SD 57501-3369
Phone: (605) 773-5032
Fax: (605) 773-3256
www.sdreadytowork.com

**Tennessee Energy Policy
Office**
Department of Economics &
    Community Development
312 8th Avenue N, 10th Floor
Nashville, Tennessee 37243
Phone: (615) 741-2994
Fax: (615) 741-5070
www.state.tn.us/ecd/energy.htm

Texas Energy Office
State Energy Conservation
    Office
P.O. Box 13528
111 E. 17th Street, Room 1114
Austin, TX 78711-3528
Phone: (512) 463-1931
Fax: (512) 475-2569
www.seco.cpa.state.tx.us/

Utah Energy Office
1594 W. North Temple, Ste 3110
P.O. Box 146100
Salt Lake City, UT 84114-6100
Phone: (801) 538-5428
Fax: (801) 538-4795
www.geology.utah.gov/sep

**Vermont Energy Efficiency
Division**
Vermont Dept. of Public Service
112 State Street
Montpelier, VT 05620-2601
Phone: (802) 828-2811
Fax: (802) 828-2342
*http://publicservice.vermont.gov/
    energy-efficiency/energy-
    efficiency.html*

Virginia Division of Energy
Virginia Department of Mines,
    Minerals & Energy
202 N. Ninth Street, 8th Floor
Richmond, VA 23219
Phone: (804) 692-3200
Fax: (804) 692-3238
www.dmme.virginia.gov

**Washington State Energy
Program**
905 Plum Street, SE, Bldg #3
P.O. Box 43165
Olympia, WA 98504-3165
Phone: (360) 956-2000
Fax: (360) 956-2217
www.energy.wsu.edu

**West Virginia Energy
Efficiency Program**
WV Development Office
Building 6, Room 553
1900 Kanawha Blvd. E.
Charleston, WV 25305
Phone: (304) 558-2234
Fax: (304) 558-0449
*www.wvdo.org/community/eep.
    html*

**Wisconsin Energy
Conservation Corporation**
Focus on Energy
431 Charmany Drive
Madison, WI 53719
Phone: (800) 762-7077
Fax: (608) 249-0339
www.focusonenergy.com

Wyoming Energy Office
Minerals, Energy &
    Transportation
Wyoming Business Council
214 West 15th Street
Cheyenne, WY 82002
Phone: (307) 777-2800
Fax: (307) 777-2837
*www.wyomingbusiness.org/
    business/energy.aspx*

Glossary

ABET. Accreditation Board for Engineering and Technology (ABET), the recognized accreditor for college and university programs in applied science, computing, engineering, and technology, made up of 28 professional and technical societies representing these fields.

AC. Alternating current; a form of electricity that cycles from minimum to maximum and then decreases to minimum again. The cycle is then repeated, now in the opposite direction. In the United States, alternating current changes direction sixty times per second.

alternative energy. Energy derived from nontraditional sources such as compressed natural gas, solar panels, hydroelectric turbines, wind turbines, and the like; such energy is alternative relative to fossil fuels, but not all forms of alternative energy (such as nuclear power) are renewable.

apprenticeship. The chief means of crafts instruction before the rise of technical and trade schools, apprenticeship is a system by which a novice serves as an assistant to a skilled worker, learning techniques and practices. Today, apprenticeship typically combines on-the-job training with supplemental reading or coursework.

architect. In the United States and Canada, a trained, accredited professional who designs buildings and often oversees their construction.

automotive designer. A specialist in the design of automobiles and other vehicles.

biobutanol. A noncorrosive combustible alcohol derived from *biomass* and used, among other purposes, as a fuel.

biochemist. A scientist who studies the chemical processes of living things, such as the production of sugars in plants.

biodiesel. Ester obtained by reacting methanol with vegetable oil, used in pure form or blended with diesel. Rapeseed, sunflower, palm, soybean, and peanut oils are most commonly used in biodiesel preparations.

bioenergy. Fuel or electricity derived from biomass. See also *ethanol*.

bioethanol. Ethyl alcohol derived from the fermentation of sugar from wheat, sugar beets, corn, and other plant sources.

biofuel. See *bioenergy*.

biogas. A liquid or gaseous fuel made from *biomass*. Some biogas is made by recovering methane from landfills, stockyards, and similar venues.

biologist. A scientist who studies living organisms and their relationship to the environment.

biomass. Organic matter available on a permanent, renewable, and sustainable basis, such as trees, agricultural crops, and waste (such as straw, wheat stalks, and pulp). The complex carbohydrates in such matter is converted to *biogas* and other forms of *bioenergy*.

blue-collar. A sociological term used to designate skilled and unskilled workers in industrial or manual jobs, so called because of the coarse blue shirts such workers wore, in contrast to administrative and managerial (white-collar) jobs.

Btu. British thermal unit; a unit of thermal energy used in reference to heating or cooling. The amount of energy required to raise the temperature of a pound of water by one degree Fahrenheit.

carbon-neutral. Carbon neutrality means balancing the amount of carbon dioxide released into the atmosphere by sequestering an equal amount. In principle, carbon neutrality is difficult for an individual to achieve; reducing CO_2 emissions through conservation has proven the more efficient tactic.

cellulosic biomass. The fibrous, woody, usually inedible portions of plants, comprising about 75 percent of all plant material and used to make biofuel.

chemical engineer. An engineer who specializes in using chemical processes in industrial uses, as in the production of biofuels from cellulosic materials.

CHP. Combined heat and power; the generation of electrical power and usable heat from a combustion process.

civil / structural engineer. An engineer who specializes in the construction of built structures such as dams, bridges, roads, canals, and buildings, in the latter often working closely with an *architect*.

DC. Direct current, a form of electricity in which the current moves in only one direction, unlike *alternating current*.

electric car. Automobile powered by electricity, rather than by petroleum or other liquid fuel.

EMCS. Energy Management Control Systems; tools that promotes energy efficiency by accurately monitoring energy consumption and building operations.

electrical engineer. An engineer who specializes in the design, construction, and maintenance of electrical systems.

energy analyst. A trained specialist charged with studying the uses and costs of energy in buildings, manufacturing processes, and the like, usually with an eye to improving efficiencies.

energy audit. An inspection process that determines how much energy is used, usually accompanied by suggestions for conserving energy.

energy efficiency. An engineering concept whereby machines, structures, manufacturing processes, and the like are designed or revised to decrease the amount of energy used, the governing idea being to improve efficiency rather than have to generate more power.

energy manager. Within industry, an engineer or trade worker whose principal task is to conduct an *energy audit*, identify ways to improve *energy efficiency*, and the like.

environmental engineer. An engineer who specializes in applying scientific and engineering principles to improving the state of the environment, often by remediating polluted sites, and who helps identify sites for the construction of utilities, roads, and other such structures.

ethanol. *Biofuel* derived from grain, corn, sugarcane, or other starchy plant material and used alone or as an additive to gasoline.

fenestrations. Glazed aperatures in buildings, including windows, glass doors, clerestories, and skylights.

fossil fuels. Carbon- and hydrogen-laden fuels formed underground from the remains of long-dead plants and animals, such as crude oil, natural gas and coal.

FSEC. Florida Solar Energy Center; a research institute of the University of Central Florida.

fuel cell. An electrochemical device that converts chemical energy to electrical energy without combustion.

fuel engineer. An engineer who specializes in the production of fuels of various kinds.

geochemist. A scientist who specializes in the study of the chemical composition of the earth and other planets.

geoexchange system. An electrically powered heating and cooling system for interior spaces, using the earth as a heat source and heat sink.

geologist. A scientist who specializes in the study of the origins, composition, and physical processes of the earth and other planets.

geophysicist. A scientist who specializes in the study of the physics of the earth, particularly its electrical, gravitational, and magnetic fields.

geothermal energy. Heat generated by natural processes within the earth. The chief energy resources are hot dry rock, magma (molten rock), hydrothermal (water/steam from geysers and fissures), and geopressure (water saturated with methane under tremendous pressure at great depths). Such heat is typically brought up through pipes with the use of a *geoexchange system*.

GHP. Geothermal heat pump; a heat pump that uses the earth as a heat source and heat sink.

gigawatt (gW). Unit of energy equivalent to one billion *watts* or one million *kilowatts*.

green building. A building or construction method that makes use of sustainable, renewable materials and principles and strives for *energy efficiency*.

green transportation. Vehicles such as bicycles or fuel cell–powered cars that do not require the burning of fossil fuels for their power.

green-collar. A play on words of the old designations blue-collar and white-collar. Refers to jobs in the "green" field of renewable energy.

HVAC. Heating, ventilation, and air conditioning, an acronym used to describe such systems and the engineers who work with them.

hybrid car. A vehicle that uses both an on-board rechargeable battery system and a fuel-based power source for propulsion. Most hybrid vehicles recharge their batteries by capturing kinetic energy from braking, with fuel used either to generate more electricity through a generator or to provide extra power when it is needed.

hydraulic engineer. A civil engineer who specializes in the study of moving fluids, particularly in how to control that flow and convey it to places where it is wanted.

hydroelectric power. Electrical power generated by the kinetic energy of moving water, often at a natural or artificial waterfall or in a swiftly moving river, but increasingly in oceanic environments through the harnessing of *wave energy*.

hydrogen energy. Energy derived from hydrogen, a renewable and emission-free fuel, now largely applied to automobiles.

hydrologist. A scientist who studies the distribution of water in the environment, often helping determine where to place dams, wells, and other structures.

hydropower. See *hydroelectric power*.

IBEW. International Brotherhood of Electrical Workers

information technology (IT). Deals with the use of electronic computers and computer software to convert, store, protect, process, transmit and retrieve information, securely.

kilowatt (kW). Unit of electrical power equal to one thousand *watts*.

LEED. Leadership in Energy and Environmental Design; a nonprofit organization that has established the Green Building Rating System and other guidelines in *sustainable building*. See *www.usgbc.org*

life-cycle engineering. An emerging research field closely related to sustainable development. LCE comprises assessment and costing models that aim at cost control, efficiency, and optimization. LCE defines the life cycle of a product from raw material through its use, disposal, and, ideally, reuse.

marine energy. See *wave energy*.

material scientist. A scientist or engineer who specializes in the study of matter and its properties and applications in industrial uses.

mechanical engineer. An engineer who specializes in the design, construction, and maintenance of mechanical systems, such as a power plant.

megawatt (MW). One million *watts*, or a thousand *kilowatts*.

meteorologist. A scientist who specializes in the study of climate and weather systems, often with an interest in forecasting near-term conditions.

methane recovery. See *biogas*.

microbiologist. A biologist specializing in the study of microorganisms, such as protozoa, fungi, bacteria, and viruses.

micro-hydro. Small-scale hydropower, producing a hundred *kilowatts* or less.

NABCEP. North American Board of Certified Energy Practitioners, an association that offers training and certification to renewable-energy professionals.

NREL. National Renewable Energy Laboratory; a research facility run by the U.S. Department of Energy in Golden, Colorado, devoted to all forms of renewable energy.

nuclear engineer. An engineer specialized in working with nuclear materials, whether in the production of *nuclear power*, in nuclear medicine, or in nuclear-weapons design and construction.

nuclear power. Energy derived from the fission of nuclear materials, producing steam and thus electricity. In a nuclear power plant, a reactor contains a core of nuclear fuel, primarily enriched uranium. When atoms of uranium fuel are hit by neutrons, they split, releasing heat.

oceanographer. A scientist whose area of study is the ocean and its environments and inhabitants.

PE. Professional Engineer; in the United States and Canada, a registered or licensed engineer authorized to offer professional services directly to the public. A PE must be a graduate of an engineering program accredited by *ABET* and have passed several rigorous examinations, among other requirements.

photovoltaic (PV). The technology of converting sunlight directly into electricity through the use of photovoltaic (solar) cells. Photovoltaic cells are made primarily of silicon, which responds electrically to the presence of sunlight, a phenomenon known as the photovoltaic effect.

process engineer. An engineer who develops industrial processes that are in turn used to make other things. Most process engineers hold degrees in chemical or mechanical engineering and have extensive knowledge in how a chemical or manufacturing plant works.

project manager. In many trades such as construction, the person in charge of establishing and maintaining schedules, budgets and payroll, staffing, and the like.

quadrillion. 1 quadrillion Btu equals 1,000,000,000,000,000 Btu (British Thermal Unit).

RA. Registered Architect, certification earned through the National Council of Architectural Registration Boards following graduation from an accredited architecture school by fulfilling a number of educational and work requirements.

renewable energy. Any form of energy that derives from sustainable resources that can be naturally renewed—wind, for instance, or energy from the sun. A general way to think of renewable energy is any energy whose production does not require burning anything, which distinguishes it from *alternative energy*.

soil scientist. A scientist who studies the chemistry and organic composition of the soil, principally in order to determine the suitability of a given plot of land to agricultural production, but also to determine siting for energy facilities.

solar collectors. Devices for capturing the sun's energy; flat-plate collectors and evacuated solar tubes are two types.

solar electricity. Electricity generated directly by sunlight via *photovoltaic cells*.

solar energy. See *solar electricity* and *solar thermal*.

solar engineer. A trained person who designs solar energy systems (solar electric systems as well as solar thermal systems).

solar installer. A trained craftsperson who specializes in installing and maintaining *photovoltaic* assemblies and other solar energy–related equipment.

solar thermal. Heat, rather than electricity, that is generated by the sun and put to a useful purpose, such as for home heating and water heating.

solar water heating. See *solar thermal*.

Stirling engine. A closed-cycle external combustion engine in which heat is applied through the wall of a chamber within which a gas is heated and cooled, thus expanding and contracting to power a piston. Stirling engines are quiet and highly efficient, but also expensive.

sustainable building. Construction that uses renewable materials and is energy-efficient, sometimes to the point of being *carbon-neutral*.

tidal energy. See *wave energy*.

turbine. A machine with propeller-like blades, or rotors, that are moved by flowing water, air, or gas to power a generator, thereby producing electricity.

utility grid. The system of electrical utilities, tied in North America into a grid such that electricity generated in California can, in theory, be transmitted to Alaska or Florida.

watt (W). Unit of electrical power used to indicate the rate of energy produced or consumed by an electrical device.

wave energy. Technically, a form of solar energy, whereby ocean waves are generated by solar-driven winds. Systems are in development to make use of this clean and endlessly renewable source of energy.

white-collar job. See *blue-collar*.

wind energy. Electrical power derived from moving air via a turbine that drives a generator.

Index

Numbers shown in italics represent photos or illustrations.

The Author

Gregory McNamee is a writer, editor, photog-
rapher, and consultant in publishing, film, and
other media. He is the author or title-page editor of twenty-six books, including several on
ecology and natural history, and of more than
three thousand periodical pieces. He writes on
science, culture, the environment, and other
topics for many publications in the United

States and abroad, and he is a contributing editor to the Encyclopaedia
Britannica and its blog. McNamee divides his time between his home
in Tucson, Arizona, and his migratory habitats in Virginia, New York,
Italy, and other places.

OTHER RENEWABLE ENERGY BOOKS BY PIXYJACK PRESS

Got Sun? Go Solar: Get Free Renewable Energy to Power Your Grid-Tied Home

Power With Nature: Alternative Energy Solutions for Homeowners

HYDROGEN—Hot Stuff Cool Science: Discover the Future of Energy

Crafting Log Homes Solar Style: An Inspiring Guide to Self-Sufficiency

To order autographed copies,
or wholesale orders for your school, library or organization,
please contact us.

PixyJack Press LLC

PO Box 149, Masonville, CO 80541 USA
www.PixyJackPress.com   info@pixyjackpress.com
A solar and wind-powered independent publisher. Members of the Green Press Initiative.